2nd International Conference on Cultural Communication and Space: Climate Change and the Built Environment

IOP Conference Series: Earth and Environmental Science
Volume 903

Denpasar, Indonesia
5 August 2021

ISBN: 978-1-7138-4885-1
ISSN: 1755-1307

Printed from e-media with permission by:

Curran Associates, Inc.
57 Morehouse Lane
Red Hook, NY 12571

Some format issues inherent in the e-media version may also appear in this print version.

Institute of Physics
Dirac House, Temple Back
Bristol BS1 6BE UK

Phone: 44 1 17 929 7481
Fax: 44 1 17 920 0979

techtracking@iop.org

Additional copies of this publication are available from:

Curran Associates, Inc.
57 Morehouse Lane
Red Hook, NY 12571 USA
Phone: 845-758-0400
Fax: 845-758-2633
Email: curran@proceedings.com
Web: www.proceedings.com

TABLE OF CONTENTS

Preface

Peer Review Declaration

Resilient Experiments in Rural Housing: Architectural Experiments in Planned Housing 1
G Galford, L M Tucker

The Correlations Between Building Heights and Wind Speed in Determining the Dimension of Windows in a High-Rise Residential Building Façade 10
S S Napitupulu, G Hardiman, RR Tobing

Composing the Noise Space: Musical Architecture of Urban Culture 18
F X T B Samodra, I G Nugrahani

Sustainable Development of Taman Harmoni Tourism Area, Karangasem Based on Local Wisdom's Value 26
N R P Salain, N M M Mahastuti

Commoning: Towards the Eviction of Kampung Bidaracina, Jakarta 41
Muhammad Luthfi Fikri, Herlily

Environmental Concern to Save Langgar Tinggi Heritage Mosque at Kampung Pekojan 53
Gianoora Achmad, Kemas Ridwan Kurniawan

Identification of the Architectural Housing Form and the Existence of Ulos Communities in Hutaraja Village, Samosir, North Sumatra, Indonesia 60
R R Tobing, A K Sakti

Privately Managed Green Open Space and Its Ecological Role in Improving Thermal Comfort of Kesiman Kertalangu Area 71
I G Edhy, N K A Siwalatri, N M Yudantini, Gusti Ayu Made Suartika

The Existence of Stilt Houses Post-Disaster 83
Andi Hildayanti, Wasilah

Local Traditions and Environmental Preservation - Genius Loci in Penglipuran Village in Protecting Public Spaces 93
I N U Mahardika, Gusti Ayu, Made Suartika, Kadek Edi Saputra

Conversion of Public Green Open Spaces to Facilitate the Tourist Industry 106
I Gusti Ayu Dwi Muliasari, Gusti Ayu Made Suartika, Kadek Edi Saputra

Road Intersection as a Public Space and Venue for Advertisements in an Urban Setting 119
I Wayan Yoga Bahagia Putra, Gusti Ayu Made Suartika, Kadek Edi Saputra

Riverfront Development - A Public Realm, Its Adaptations, and Environmental Impacts 134
Luh Made Ratih Ratna Dewi, Gusti Ayu Made Suartika, Kadek Edi Saputra

Carrying Capacity of Land in Bali Based on Ecological Footprint 147
A A B Bayu Baskhara Hutama, A K Muktiwibowo, Widiastuti, K E Saputra

Author Index

IOP Conf. Series: Earth and Environmental Science **903** (2021) 011001 doi:10.1088/1755-1315/903/1/011001

2nd International Conference on Cultural Communication & Space (ICCCS)

CLIMATE CHANGE AND THE BUILT ENVIRONMENT

Level 1, GDLN Building, Udayana University Denpasar Campus
Jl. P.B. Sudirman, Denpasar - Bali (80232), Indonesia
5 August 2021

https://icccs.unud.ac.id

PREFACE

Gusti Ayu Made Suartika

In 1896 the Swedish scientist and Noble Prize winner Svante Arrhenius showed that a doubling of atmospheric carbon dioxide would result on earth in increased surface temperature of 5-6 degrees Celsius. His warning has largely gone unheeded for the last 124 years, and we are already in an era of disaster management now guaranteed to get significantly worse. The response of nature is ubiquitous and terrifying-prolonged drought, bushfires, extreme flooding, unbearable heat waves, freezing storms, rising sea water levels, melting polar icecaps and glaciers, coastal abrasion, and other phenomena. Some believe these conditions are natural - the adopted manner by which the earth chooses to rejuvenate itself and therefore to deny we have any responsibility for its effects. Climate change sceptics promote 'fake news' largely to perpetuate business interests for a few more years, in a lemming like rush towards disaster. Or as O'Connor's second law states, 'capitalism always destroys the source of its own success.'

Nature's responses to centuries of abuse do not however address the human cost of global warming – everything from health issues, increasing costs of social services, the loss of livelihood and property, and the potential adaptation of our entire built environment in the interests of survival. The 2nd International Conference on Cultural Communication and Space (ICCCS) promotes the viewpoint that concerted action is needed across a wide range of offensives – providing alternative energy sources; remodelling the built environment; reducing both personal and industrial consumption of nature, radically changing political systems, an immediate triple bottom line approach to all business, and a sceptical view of technology as having any capacity to ameliorate the above problems. Even then a sustainable planet might not result.

While many academic events have discussed these matters across the whole spectrum of problems, we focus here on the built environment in relation to climate change phenomena and its required responses to culture and space. Given the above

IOP Conf. Series: Earth and Environmental Science **903** (2021) 011001 doi:10.1088/1755-1315/903/1/011001

considerations, this second ICCCS focuses on the following topics of conversation that address the significant policies and practices of climate change:

1. Implications for built form and environment.
2. Innovative design and construction for eco-urbanism.
3. Restructuring development & planning.
4. Political, socio-cultural and behavioural reconstitution.

We hope that in this conference, imagination, risk-taking and creative solutions will be proposed that reflect responses as radical as the problems. We no longer live in an era where 'political correctness' in regard to climate change can be supported. Without risk there can be no forward movement. Or as Ray Bradbury said 'Living at risk is jumping off the cliff and building your wings on the way down.' We owe the planet as big a jump as we can make.

This 2021's conference is organized following the 1st ICCCS held in 2018. Since it is a biannual event, it was first scheduled for 26-27 November 2020. But, due to the Covid-19 pandemic, the organizer decided to reschedule the gathering to 5 August 2021 with a hope the pandemic would be over and it would be possible to hold the 2nd ICCCS in a face-to-face mode. Unfortunately, the pandemic had not shown any sign of stopping leaving the organizer with no choice but to organize it as an online event following the schedule. As expected, this had a dire implication especially in regard to the number of participants presenting their research studies. In comparison to the 1st ICCCS, this number significantly dropped from 97 to 14. This is a reduction of almost 85%. Let us all hope that the before-Covid-19 pandemic life will return soon. Thus, it will no longer be of a deterrence in bringing the audience back in the upcoming 3rd ICCCS.

Gusti Ayu Made Suartika

IOP Conf. Series: Earth and Environmental Science **903** (2021) 011001 doi:10.1088/1755-1315/903/1/011001

CONFERENCE INFORMATION

Conferences tracks:

1. Climate change and its implications on the built forms and environment
2. Innovative design and construction in response to climate change
3. Development & planning for climate change
4. Coping with climate change: political, socio-cultural and behavioural perspectives

Conference dates and venue

The Conference was held on:

- Date : 05 August 2021
- Time : 08.00 am – 17.00 (Bali time - GMT + 8)
- Venue : Level 2- GDLN Building of Udayana University Denpasar Campus, Jl P.B.Sudirman, Denpasar-Bali (80232), Indonesia

Speakers

- Keynote speakers:
  1. Prof. Alexander Cuthbert (University of New South Wales Australia)
  2. Dr. Ken Yeang (Architect, Ecologist, Planner, and Author, Malaysia)

Scientific Committee

1. Prof. Joseph Aranha, USA
2. Prof. Martin Loosemore, Australia
3. Prof. Alexander Cuthbert, Australia
4. Dr. Graeme McRae, New Zealand
5. Dr. Campbell Drake, Australia
6. Dr. Mark Titmarsh, Australia
7. Prof. Abidin Kusno, Canada
8. Prof. Gunawan Tjahjono, Indonesia
9. Assc.Prof. Ranjith Dayaratne, Bahrain
10. Prof. Pratyush Shankar, India
11. Prof. Yu Liu, China
12. Dr. Heru Wibowo Poerbo, Indonesia
13. Prof. Deden Rukmana, USA
14. Dr. Mark Russell, USA
15. Dr Ahmad Gamal, Indonesia
16. Dr. N K P. Dewi Jayanti, Czech Republic
17. Dr. Himasari Hanan, Indonesia
18. Indah Widiastuti, PhD, Indonesia
19. Dr. Cut Dewi, Indonesia
20. Dr. Maria Immaculata Ririk Winandari, Indonesia
21. Punto Wijayanto, Indonesia
22. Dr. Dwita Hadi Rahmi, Indonesia

IOP Conf. Series: Earth and Environmental Science **903** (2021) 011001 doi:10.1088/1755-1315/903/1/011001

23. Prof. Gusti Ayu Made Suartika, Indonesia
24. I Nyoman Gede Maha Putra, PhD, Indonesia
25. Ulli Herbig, PhD, Austria
26. Prof. Ramesh Biswas, Austria
27. Christine Cabasset, PhD, France
28. Yohannes Firzal, PhD, Indonesia
29. Gun Faisal, Indonesia
30. Dr. Agus Suharjono Ekomadyo, Indonesia
31. Dr. Tri Anggraini Prajnawrdhi, Indonesia
32. Dr. Ni Made Yudantini, Indonesia
33. Dr. I Dewa Gede Agung Diasana Putra, Indonesia
34. Dr. Karel Muktiwibowo, Indonesia
35. Dr. Ngakan Ketut Acwin Dwijendra, Indonesia

Conference timeline and deadlines

1. Abstract submission	07 April 2021
2. Abstract acceptance	10 April 2021
3. Full paper submission	04 June 2021
4. Review result for full paper	20 June 2021
5. Submission for revised full paper	20 July 2021

Publications

1. IOP – Earth and Environmental Science Conference Proceedings (Scopus indexed Conference Series)

Participants

The 2nd ICCCS engaged academics, researchers, government officers, non-government organizations/individuals as well professionals of relevant field and expertise. Participations in this conference were accommodated in four different forms, which were as follows:

- Paper publication and oral presentation
- Paper publication and poster presentation
- Academic conference paper publication without presentation
- Participants without presentation

Secretariat

Cultural Communication and Space Research Centre, Udayana University Denpasar Campus, Jl. P.B. Sudirman, Denpasar-Bali (80232) - Indonesia.
Email : icccs@unud.ac.id

Peer review declaration

All papers published in this volume of IOP Conference Series: Earth and Environmental Science have been peer reviewed through processes administered by the Editors. Reviews were conducted by expert referees to the professional and scientific standards expected of a proceedings journal published by IOP Publishing.

- **Type of peer review:** Double-blind
- **Conference submission management system:**
  - https://ucs.unud.ac.id/conf/icccs
  - Email: icccs@unud.ac.id
- **Number of submissions received:** 36
- **Number of submissions sent for review:** 36
- **Number of submissions accepted:** 14
- **Acceptance Rate (Number of Submissions Accepted / Number of Submissions Received X 100): (14/36) x 100%** = 38,89%
- **Average number of reviews per paper:** 2
- **Total number of reviewers involved:** 35
- **Any additional info on review process: -**
- **Contact person for queries:**
  - Name : Gusti Ayu Made Suartika
  - Affiliation : Udayana University-Bli, Indonesia
  - Email : ayusuartika@unud.ac.id

Resilient Experiments in Rural Housing: Architectural Experiments in Planned Housing

G Galford[1] and L M Tucker[2]

[1] Assistant Professor, Residential Environments + Design, Virginia Tech, Blacksburg Virginia USA
[2] Professor, Interior Design Program, School of Architecture + Design, Virginia Tech, Blacksburg Virginia USA

Email: ggalford@vt.edu

Abstract. Single family houses contribute substantially to climate change in the US and other parts of the world. In the US specifically, most housing has been designed by builders and developers. The motivation has not been sustainability and a knowledge of how to design net zero energy and net zero water dwelling is not commonly understood. This paper seeks to use a historical model as viewed through the lens of the Living Building Challenge to demonstrate how an architect designed historic example might provide a way of implementing a cutting-edge approach to sustainable housing today. Arthurdale was an early 20th century housing experiment that was conceptualized to provide for sustainable living in rural Appalachia. This paper presents the history of the region, an overview of the houses and the Living Building Challenge and then analyses how this historic prototype might model a sustainable housing development today using the Living Building Challenge system.

Keywords: sustainable housing, living building challenge, Appalachia, new deal

1. Introduction

According to the United States Green Building Council, buildings contribute significantly to global warming and climate change. "There's a consensus among the international scientific community that human activity is a driver of global warming. There's over a 95% probability that human actions over the past 50 years have warmed our planet, said the Intergovernmental Panel on Climate Change in its Fifth Assessment Report" [1]. Buildings and their construction together account for 39 percent of energy-related carbon dioxide emissions and 36 percent of global energy use annually, according to the United Nations Environment Program making building a significant contributor to global warming [1]. Further, in the US, commercial buildings use only 40% of all energy and residential construction is responsible for the other 60% [2].As such, designing sustainable housing can have a major impact on climate change. In the US, most residential buildings are not designed by architects who are trying to impact climate change through building design. As such, committing to the Architecture 2030 challenge to eliminate the use of fossil fuels in buildings and the resulting work will impact fewer than half of all buildings. The sustainable design of residential buildings is critical to the reduction of greenhouse gasses and climate change.

The design of houses in the US has primarily been done by the construction industry, most specifically by residential property developers. For a variety of reasons (capitalism, lack of financial

IOP Conf. Series: Earth and Environmental Science **903** (2021) 012001 doi:10.1088/1755-1315/903/1/012001

incentive, lack of interest by architects in houses for the middle class) house design has taken place outside of the domain of designers trained, educated and licensed to work for the public. This has led to a proliferation of large, inefficient, and mostly unsustainable residential construction. Fortunately, there are several historic examples of efforts to try and rectify this issue. In the early 20th century beginning with U.S. groups such as the Architects' Small House Service Bureau in the 1920s and 30s and followed by programs such as the federal subsistence homestead projects, architects and politicians focused on the issue of well-designed and affordable houses. With these programs, trained designers and architects produced house plans and kit houses that could be adapted to different areas of the country, sited to take advantage of local sun and wind patterns, and which could have gardens on the property to grown one's own food combined with root cellars and other sustainable features [3]. Using local materials and construction techniques, houses were sited to optimize the plot of land upon which they were built. Well-designed and efficient floor plans led to smaller houses and the use of less materials. Windows and doors planned for cross-ventilation resulted in comfortable interiors throughout the year.

This time period provides the best example of concerted efforts by the design community to improve housing for most people. Architects in the US have traditionally focused on high-end single-family houses for wealthy clients or on housing developments in urban areas as utopian visions for bettering the conditions for living among the lower class. Many of these visionary projects have failed. The living conditions of the majority of the population is not accounted for in this bifurcated approach. As such, the buildings are designed, constructed and driven by capitalism and financial gain and not by issues of efficiency, design, wellness of occupants, or sustainability.

It is theorized that this residential experiment in sustainable housing was not economically successful at the time, and was seen as a failure due to that lack of quick economic payback. It was a quest to use the powers of government to provide quick societal change. It needs to be viewed within a longitudinal framework that views it success over time. Critically utilizing its unique Appalachian vernacular architecture will provide insight into its success as well. This initial research is the beginning to a larger eventual study.

1.1 The Living Building Challenge

Of the current sustainability standards and guidelines, the most visionary platform comes from the International Living Building Institute with the Living Building Challenge for buildings and communities. The program is structured around seven areas (called petals after the idea that a building should grow in place like a flower). The seven petals include: Energy, Water, Place, Materials, Beauty, Health and Wellness and Equity. Like many green building programs, there is a focus on materials, energy and water. Where the LBC goes beyond typical sustainable building approaches is in the inclusion of design for Equity, Human Health, and climactic wellness. When first developed, the LBC was viewed as nearly impossible to achieve, but it had the impact of pushing industry partners, manufacturers, and regulatory bodies to accommodate the possibility of buildings that have both net zero energy use and net zero water use. Thus, a building can exist without taking anything from its surroundings and in fact can contribute to the health of the people and place. What was visionary is now possible to achieve.

1.2 Sustainable Housing and Vernacular Strategies

Since houses were first constructed in North America, vernacular strategies tended to be sustainable prior to the Industrial Revolution [4]. As with many parts of the world, the buildings and homeowners in America relied on locally available materials, cross ventilation, and preferable siting to take advantage of prevailing winds and to harvest daylight without creating solar gain in the interiors. An understanding of how to reduce heat build-up by raising a building off the ground or providing a central courtyard are regional variations on these traditions. In the eastern US, mountainous terrain allowed builders to use local wood, stone and even make bricks on site to construct houses. Slate was also widely available as

IOP Conf. Series: Earth and Environmental Science **903** (2021) 012001 doi:10.1088/1755-1315/903/1/012001

a roofing material as were wood shakes and other materials grown locally. Oftentimes, the front door would face south to allow the long side of the house to take advantage of sun in the winter and would be shaded by deciduous trees in the summer months. This prevented solar heat gain in the morning or evening on the east and west ends.

With the industrial revolution, many traditional ways of building and the use of local materials was supplanted by widely available manufactured materials first by rail and later through interstate trucking. New building systems such as indoor plumbing, electricity, and central heating transformed the American home [4]. More recently, a return to vernacular methods has been embraced as the urgent need to confront climate change has infiltrated the design and construction industries. The new Urbanism movement sought to recreate some of the neighbourhood connections of turn of the century developments. Specific Green standards now address single family houses in North America for example, LEED Homes and New Development Green Building Rating Systems, the Living Building Challenge (both for individual buildings and developments), PassivHaus standards and others. The need to address house design has been recognized but the path forward is less than clear as most houses are not designed by the professionals embracing these systems.

Having a series of model homes designed to meet sustainability criteria with specifications and siting adaptations have been proposed historically (Architects' Small House Service Bureau 1920s-1940s) and the Homesteader's Program in the 1940s. A modern version of this approach can be informed by these historic programs. When viewed through the lens of the LBC, the Arthurdale Appalachian housing project provides an historic example of how to achieve many of the aims of a sustainable housing stock.

2. Literature Review

2.1 This History of Arthurdale

The history of the region of Appalachia is unique within American history. The mountain range which defines its boundaries stretches from Georgia to New York [5]. This work focuses on areas closer to its centre, within the state of West Virginia. Early American colonial history saw these mountains as a natural barrier to settlement, providing a barricade that often-defined Native American lands [6]. The desire for new lands for settlement eventually brought incursion into the mountains with the original farming settlers being of English, German or Scots-Irish descent. Resiliency was required for these early settlers as outside support was minimal.

The Industrial Revolution of the late 19th century produced a need for raw materials to supply manufacturing facilities as well as to provide heat for burgeoning cities. The discovery of coal within the Appalachian Mountains, in an addition to its vast forests of timber, brought outside speculators into the region to begin a long process of land speculation [7] as underground mineral rights were sold by small farm land owners [8]. Ownership of land in Central Appalachia transitioned from a model of small landowner independence to one largely controlled by outside industrial speculators with little or no connection to area residents. From residential construction to steel production, the natural and human resources of Appalachia fuelled the American economy of the Industrial Revolution.

The region truly opened up with the construction of railroads into the mountains in the late nineteenth century [9]. The railroads were able to connect the population centres of the Eastern seaboard with the settlements of the Ohio River Valley and mid-western cities beyond. Providing access to Central Appalachia most importantly provided the means to move the timber and the coal from the mountains to the industrial cities that encircle the region [10]. The steel production that helped build the nation would not have occurred without its proximity to Appalachian coal [11].

There was not sufficient labour to do the necessary work, so outside coal and timber companies recruited newcomers to the area, whether African Americans from southern states, or new immigrants from Europe. This influx of labour to the mountains from distant regions re-shaped the culture of the region, in settlements planned, owned, and controlled by coal company operators [12]. These towns were connected to the mines they served, with necessary public services supplied by the coal company [13]. Schools, clinics, churches, department stores, recreational centres and even cultural venues were incorporated into the original towns to attract and retain workers. Miners had to rent housing versus own their homes [14], and their pay was in company scrip, ensuring that the miner and his family were economically bound together with their employers [15]. Due to the rugged terrain, these towns were often isolated from each other, and contact with other areas of the country was difficult. Original mines and timber concerns often only housed workers, but families were soon included to maintain employee contentment. Hierarchies within these towns were often evident through the size of housing and location. Manager and owner housing would be of a larger type, and would be situated higher up the mountain side so that they would be able to observe workers from above. During the greatest expansion of these mines, these towns could be opulent in their civic architecture. Racial segregation was common in these towns, reflecting norms within the country [16].

Life within the mines in the late nineteenth and early twentieth centuries was indescribably hard. Men worked extremely long hours underground in dangerous conditions where a methane explosion or a roof collapse was not uncommon. Mine deaths were frequent. Women and children were dependent on the mine owners for both a livelihood and for housing, and when a husband was killed in a work accident, they often had to vacate as the town was only designed for those who contributed to its economic mission [17]. With time, labour movements in the rest of the country affected the coal fields and produced unionization movements [18]. Appalachian history is replete with violent battles between coal miners and mine operators which ultimately reduced the power of the United Mine Workers Association [19]. This eventually brought about improved living conditions, but the 1950s saw an increase in mechanization of the mines which reduced labour needs [20]. This has continued and the mining labour force in the region is presently a fraction of its peak. Strip mining and mountain top removal have left permanent environmental damage, as coal mining in the region is now affected by renewable energy sources that are now economically viable.

The coal company towns of the 1930s were deeply affected by the Great Depression as First Lady Eleanor Roosevelt visited towns devastated by the economic downturn [21]. One mining town she visited was Scott's Run, West Virginia [22]. Many residents were unemployed due to labour stoppages and conditions that were deemed deplorable [23]. The economic devastation in the coal fields was overwhelming, so she and other government planners used the opportunity to experiment with three new towns in the state that would be viewed as government-owned subsistence homesteads, meaning that they would be sustainable for the families who relocated there [24]. These communities had houses with indoor plumbing, electricity, sufficient land for gardening and food producing animals, and an underground root cellar. The most well-known of the three towns is Arthurdale in Preston County, WV. The town of 165 homes was comprised of two major types. The first was a prefabricated type from the E.F. Hodgson Co, that was designed for more seasonal use, and not for cold mountain winters. The second type was built locally and synchronized better with the climate [25]. This town also maintained its own school, a community building for public events, an artisan workshop for the production of woven goods and furniture, and a small nearby factory for the production of household appliances.

The initial costs of the housing for the community was more than anticipated and members of the US Congress were opposed to strong government intervention in the housing industry. The US federal government sold the town in 1946 to private owners, ending the governmental experiment. Arthurdale, as well as the other two towns in Randolph and Putnam counties in West Virginia, are all active communities today.

3. Methods

This study uses a case study approach [26] where Arthurdale is an example of an historic housing project that can inform the implementation of the Living Building Challenge. A comparison of the LBC to features of the Arthurdale homes demonstrates which sustainable features can be found in this type of vernacular architecture in Appalachia. Table 1 compares the key components of the Living Building Challenge (LBC) with the features of the house designs for Arthurdale.

Table 1. Comparison of Arthurdale Homes and LBC model

Living Building Challenge	**Arthurdale Houses (Hodgson, Wagner and Stone)**	**Notes**
PLACE		
01. Ecology of Place	Locally adapted to site	
02. Urban Agriculture	On-site agriculture (and livestock)	Cows, pigs and chickens
03. Habitat Exchange		Not applicable at the time
04. Human-Scaled Living	Porches	
WATER		
05. Responsible Water Use	Rain water collection?	Would need adaptation
06. Net Positive Water		
ENERGY		
07. Energy + Carbon Reduction	Double walls were designed with "arctic" lined Celotex insulation (1/2" cane fibre insulation); doors were fitted with screens to permit natural ventilation	Would need adaptation
08. Net Positive Carbon		No applicable at the time
HEALTH + HAPPINESS		
09. Healthy Interior Environment	Kitchen were designed to be of the latest "sanitary" technology. They were designed to be both cheerful (full of light) and convenient; sunrooms had plenty of natural light	
10. Healthy Interior Performance	Interiors were designed to evoke a sense of comfort and this was used in the advertising by Hodgson Company	
11. Access to Nature	Windows all rooms Sited for growing food and raising animals on the property	
MATERIALS		
12. Responsible Materials	Locally available Prefabricated panel construction used in phase 1	
13. Red List		Not applicable at the time/could be used
14. Responsible Sourcing	Locally available	
15. Living Economy Sourcing	Local workmanship/created jobs	
16. Net Positive Waste	Use of prefabricated construction resulted in a lack on construction waste on site	
EQUITY		
17. Universal Access	Most houses were located on a single level. Although ADA was not yet law, this would have contributed to a more accessible interior	
18. Inclusion	Economically equitable	

IOP Conf. Series: Earth and Environmental Science **903** (2021) 012001 doi:10.1088/1755-1315/903/1/012001

BEAUTY		
19. Beauty + Biophilia	Houses were designed with plenty of natural daylight and windows for views. Lots were sized to accommodate growing food and other plants on the site.	
20. Inspiration + Education	The development relied on the Homesteading program as an educational tool for residents.	

4. Discussion

Historic housing projects can provide examples for how to take the complexity of the Living Building Challenge and scale it to a rural housing development. Instead of the typical approach US architects use towards designing one house at a time, these historic housing developments can provide a scalable model for implementation of sustainable housing for an entire neighbourhood development. Informed by good design practices focused on sustainability provided by the LBC and combined with historic large-scale developments both sustainability and scale can be achieved, which can inform both designers and residential developers. The Living Building Challenge also includes a provision for 'scale jumping' which allows for net zero energy and net zero water to be achieved at the level of neighbourhoods. This includes rain water catchment and on-site wetland water processing in combination with renewable energy strategies such as solar and wind collection that can fuel and provide water for all houses in the neighbourhood without being located discretely on each individual house.

The challenges faced in Appalachia related to climate change included the clear cutting of forests (ca. 1910) which resulted in flooding, fire and deforestation on a massive scale as well as the use of coal, the primary local industry [27]. Both of these can be mitigated through the LBC lens. For example, collecting rainwater and either providing on-site treatment to convert it to potable water or using it as grey water for toilet flushing are a way to prevent flooding. Xeriscape landscaping approaches wherein local plants are used to help absorb excess rain water are another solution. A neighbourhood solar array or wind turbines could supplement rooftop photovoltaics to provide 100% renewable energy stores for the houses. Treating the entire development as a living organism that is self-supporting would eradicate the ill-effects of climate change caused by the buildings themselves (both the construction and maintenance of them).

The housing for Arthurdale is an example of a political movement to "back to the land" which started with Ebenezer Howard's Garden City concept and was extended to veterans in the late 19th century. Following the Depression, enabling legislation from 1933 sought to redistribute the population across the US. In response to an outpouring of letters to the President, Eleanor Roosevelt chose to take up the fight for miners in Appalachia and of the over 400 proposals, Arthurdale Sustenance Community was the project chosen for support [28]. The proposal included houses, community building and support systems for the residents of the area. *Each house had its own area for growing food, a place for a cow (which was free to each resident) and a root cellar.*

The houses in Arthurdale were built in phases. The first phase was based on a prefabricated modular unit by E.F. Hodgson Company. Modular construction saves on waste and is inherently sustainable as the panels are built to conventional material sizes. The first phase of houses in Arthurdale was the topic of an article in the *Saturday Evening Post* August 4, 1934. One of the Hodgson designs was shown in the article.

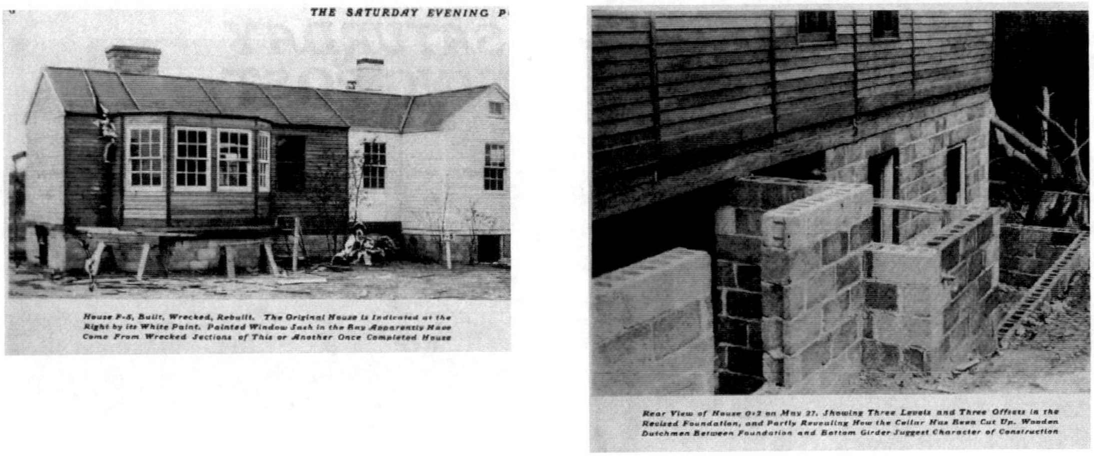

Figure 1. Image from Saturday Evening Post article August 4, 1934, p. 6.

The seams from the pre-constructed modular panels of the walls and roof are clearly visible in the photograph. Modular house design is something many architects focus on for sustainable construction today. According to S. Allen Chambers for Archipedia, the Hodgson Houses were built in shapes for which they were named: L, H, T, and I shaped. Fifty Hodgson houses were constructed of which 48 remain [29]. Each building site had one to four acres which included a combination cow shed/chicken house, vegetable garden space and flower gardens as well as a place for an orchard. Later phases were constructed on site. The later phases looked very different from the Hodgson houses. They were largely constructed from stone. The second phase consisted of 75 additional houses. The Wagner houses (named after the architect who designed them) were 1 ½ stories and either made from local lumber or stone set atop a concrete foundation. The third and final phase included 40 additional houses and were designed in one of three popular styles—Tudor Revival, Colonial Revival or 1 ½ story bungalows. A total of 165 houses comprised the final Arthurdale project which was completed in 1937 [29].

The Arthurdale project was not an economic success [30]. The houses cost twice as much as other comparable houses at the time. The community-based design concept was viewed by some as communist, and the overall approach was eventually abandoned and the development was sold. Despite this, this experiment represents one of the only times a holistic program for the poor based on sustenance housing was ever conceived and executed before or since in the US [30].

5. Conclusions

This paper demonstrates how a historic development (sustainable at its time in history) can provide a model for how to approach climate change in the house design and construction industry today. Although this work is an overview of the sustainable features that were integral to the Arthurdale houses, future work will focus on a mixed methods study that includes an analysis of current houses that remain from the different phases of construction and how they have been modified over time. Additionally, current and previous residents will be surveyed with some focus interviews. The three-part approach to sustainability—social, economic, and environmental—are all aspects contained within the Arthurdale project.

Works on 'green' housing often review vernacular architecture to learn lessons regarding construction techniques, but this work is focused on sustainability when performed by governments planning at a community scale. This work is unique in that it has a significant longitudinal advantage in how it can view the town, while using a modern system of sustainability to evaluate its success or failure.

IOP Conf. Series: Earth and Environmental Science **903** (2021) 012001 doi:10.1088/1755-1315/903/1/012001

This work shows that current systems can evaluate historical models and can be applied to similar situations.

Future work will include a mixed methods study that will evaluate the current homes, from different phases of construction, and how they have been modified over time. Current and previous residents will be surveyed with limited focus interviews for the collection of qualitative data. Each successive phase will inform the next one. The goal is to provide deep evaluations of planned rural communities. The three-part approach to sustainability, social, economic, and environmental, are all aspects contained within the Arthurdale project.

6. References

[1] C. Huynh, 01 march 2021. [Online]. Available: https://www.usgbc.org/articles/how-green-buildings-can-help-fight-climate-change.

[2] D. Budds, "How do Buildings Contribute to Climate Change?," [Online]. Available: https://archive.curbed.com/2019/9/19/20874234/buildings-carbon-emissions-climate-change.

[3] P. Tedesco and J. Tedesco, "Portable and Prefabricated Houses of the Thirties: The E.F," in *Hodgson Company 1935 and 1939 Catalogs*, Boston, JBT Publishing, 2007, pp. pp. vi-ix.

[4] R. Banham, "Architecture of the Well-Tempered Environment," London, The Architectural Press, 1984, p. 24.

[5] J. Williams, "Appalachia: A history," Chapel Hill, UNC Press), 2002, p. 13.

[6] J. Williams, "Appalachia: A history," Chapel Hill, UNC Press, 2002, p. 25.

[7] H. Caudill, "Night comes to the Cumberlands: A biography of a depressed area," New York, Little Brown, 1963, p. 79.

[8] J. Williams, "Appalachia: A history," Chapel Hill, UNC Press, 2002 , pp. 247-250.

[9] R. Drake, "A history of Appalachia," Lexington, University Press of KY, 2001, p. 141.

[10] H. Caudill, "Night comes to the Cumberlands: A biography of a depressed area," New York, Little Brown, 1963, p. 74.

[11] J. Williams, " Appalachia: A history," Chapel Hill, UNC Press, 2002, pp. 242-244.

[12] J. Williams, "Appalachia: A cultural history," Chapel Hill, UNC Press, 2002, p. 259.

[13] C. Shifflett, "Coal Towns: Life, work, an culture in company towns of Southern Appalachia 1880-1960," Knoxville, University of Tennessee Press, 1991, pp. 143-146.

[14] H. Green, " The company town: The industrial Edens and satanic mills that shaped the American economy," New York, Basic Books, 2011, p. 57.

[15] R. Drake, "A history of Appalachia," Lexington, University Press of Kentucky, 2001, p. 148.

[16] H. Caudill, "Night comes to the Cumberlands: A biography of a depressed area," New York, Little Brown, 1963, pp. 102,105.

[17] H. Caudill, "Night comes to the Cumberlands: A biography of a depressed area," New York, Little Brown, 1963 , p. 121.

[18] J. Williams, " Appalachia: A cultural history," Chapel Hill, UNC Press, 2002, p. 271.

[19] J. Thomas, " An Appalachian New Deal: West Virginia in the Great Depression," Lexington, University Press of Kentucky, 1998, p. 9.

[20] H. Caudill, "Night comes to the Cumberlands: A biography of a depressed area," New York, Little Brown, 1963, p. 242.

[21] J. Thomas, "An Appalachian New Deal: West Virginia in the Great Depression," Lexington, University Press of Kentucky, 1998, p. 133.

[22] C. Maloney, "Back to the land: Arthurdale, FDR's New Deal, and the costs of economic planning," Hoboken, Wiley, 2011, p. 23.

[23] J. Thomas, "An Appalachian New Deal: West Virginia in the Great Depression," Lexington, University Press of Kentucky, 1998, p. 163.

[24] J. Williams, "Appalachia: A cultural history," Chapel Hill, UNC Press, 2002, p. 286.

[25] P. Tedesco and J. Tedesco, "Portable and Prefabricated Houses of the Thirties: The E.F. Hodgson Company 1935 and 1939 Catalogs," Boston, JBT Publishing, 2007, pp. vi-ix.

[26] D. Groat, "Architectural Research Methods, Second Edition," Hoboken, Wiley, 2013 , p. 418.

[27] H. Caudill, "Night comes to the Cumberlands: A biography of a depressed area," New York, Little Brown, 1963 , p. 68.

[28] E. Lynn Miller, "“Homesteading FDR Style at Arthurdale,” Landscape Architecture Magazine Vol. 68, No. 5," September 1978. [Online]. Available: https://www.jstor.org/stable/44666542. [Accessed 21 April 2021].

[29] S. A. Chambers, "“Arthurdale,”," [Online]. Available: https://sah-archipedia.org/essays/WV-01-0007-0002-0006. [Accessed 28 April 2021].

[30] C. Maloney, "Back to the land: Arthurdale, FDR’s New Deal, and the costs of economic planning," Hoboken, Wiley, 2011, pp. 194-195.

The Correlations between Building Heights and Wind Speed in Determining the Dimension of Windows in a High-rise Residential Building Façade

SS Napitupulu[1], G Hardiman[2], and RR Tobing[3]

[1]Doctoral Student of Architecture of Universitas Katolik Parahyangan
[2]Department of Architecture of Universitas Diponegoro
[3]Department of Architecture of Universitas Katolik Parahyangan

salnapitt.3@gmail.com

Abstract. Climate change due to architecture occurs as a result of technological developments that support the development of materials, electrical mechanics, structures, and building shapes that play a role in increasing emission levels in the air. One type of building in Indonesia that contributes to increasing emissions is the residential building known as *rumah susun*. This research employs the case study method, observing the *Rumah Susun* Jatinegara Barat, located in East Jakarta. The case study shows that the use of prototypes that are not environmentally friendly makes a building's performance worse. The use of precast, which resulted in monotonous window dimensions, is considered the main factor causing the failure of this *Rumah Susun* Jatinegara Barat to adapt to the surrounding environment. This problem occurred because the openings in the building façades had a monotonous dimension while the wind intensity that hit the building was increasing. The final result shows that the windows on the façades of the case study were not functioning except if all the openings in the residential unit are open. These results prove that the height of a building is an essential factor in planning high-rise flats, especially in Jakarta.

Keywords: climate change, passive ventilation, *rumah* susun, wind speed, window dimension

1. Introduction

Architecture is one of the main factors increasing emissions in the air and causing climate change [1][2]. This influence is inseparable from technological developments in the 19th century [3][4]. The benefits of technology providing convenience to humans make technology in architecture develop and spread rapidly. Technological advances also influence four architectural elements: materials, electrical mechanics, building structures, and building shapes. In practice, the four elements are divided into two groups, namely active and passive. Active technology belongs to the realm of materials and electrical mechanics, while passive technology can be found in the structure and shapes of the building [5][6].

The advancement of technology in Indonesia stems from the distribution of infrastructure, causing the urbanization rate to increase rapidly. This phenomenon makes large urban areas such as Jakarta become densely populated areas. In 1985, Jakarta was expanded to the borders, turning Jakarta into an overcapacity city [7]. The increasing amount of urbanization eventually transforms residential architecture in urban areas into vertical housings divided into two types, namely apartments and flats.

Apartments and flats have the same meaning: the arrangement of residential units to form a shared residence. However, if the arrangement is seen based on the four architectural elements, it can be seen that the materials, electrical mechanics, structure, and shape of an apartment building are easier to adjust because the construction cost is much higher than that of a flat. Therefore, when viewed through the relationship between climate change due to architecture and residential functions, flats have more complex problems.

Flat problems arise due to two factors, namely the planning system and the occupants' economic condition. Flat planning limits construction cost because, in contrast to apartments managed by private sectors, flats are managed by the state. This management division makes the procurement of flats focus on practical systems that aim to speed up construction time, namely the prototype system. Flat prototypes have the following characteristics: (1) the use of precast walls, (2) an asymmetrical building plan, (3) I-shaped dominant building mass configuration, (4) a monotonous window type, and (5) a unit area of 36 m^2 to 45 m^2 [8][9]. Prototypes are the leading cause of climate and architectural problems because prototype planning consists of human activities. The use of the prototype resulted in poor building performance, making the building unable to act as an intermediary between humans and nature.

The economic factor is one factor that can show the inability of a building to work correctly. Flat buildings have a low economic community target (MBR), so the ability of the building to act as an intermediary between nature (climate) and humans through windows on the building façade plays an essential role because not all flat residents can afford ventilation tools such as air conditioners (AC). However, since the flat prototypes are not environmentally friendly, the windows on the building façades cannot function optimally. This occurs because the dimensions of the windows are monotonous when the wind intensity (especially wind speed) that hits the building does not have a fixed intensity, making the use of mechanical technology inevitable. Based on GBCI data, windows on the façade play a 60% role in the successful performance of the building [10][11]. Therefore, the calculation of window dimensions on the building façade is essential to working optimally as an intermediary between nature and humans. In addition, flat buildings can become one with nature and contribute to suppressing the increasing temperature of the earth.

2. Methods

This research employs a case study method with the *rumah susun* Jatinegara Barat as the selected object in East Jakarta by focusing on the climate element of wind speed. The research object has limitations such as the character of the location, environmental conditions, and the existing conditions of the flat buildings. Some of the criteria include: (1) having a height of more than ten floors, (2) having poor window conditions, (3) being located in a river area, (4) having an I-shaped building mass, and (5) having a residential unit with a minimum area of 36 m^2, while the climate scope has been determined by the instability of motion and wind speed of urban areas if it was at the minimum altitude of 10 meters [12]. The study used a field survey to collect existing data, include room height, room area, and wall dimensions. The other data includes the thermal data of the room, namely wind speed, humidity, and room temperature. Values of humidity and room temperature units used as control variables validated the final results of the analysis. The two-field data were subsequently analyzed based on the three theories, namely by Terry S. Boutet, Givoni, and Olgyay, and technical standards in Indonesia, namely the Indonesian National Standard. With these research steps, the discussion in the paper includes understanding the theory of wind in architecture, *rumah susun* Jatinegara Barat as research objects, and field data analysis.

3. Results and Discussions

A basic understanding of the relationship between wind speed, flats, and windows is related to the intensity formed through the movement of natural elements, namely the sun and wind. The sun dominantly affects temperature through radiation transmitted by light, while wind dominantly affects humidity. In contrast to the two thermal units, wind speed intensity depends on the sun and gravity. The higher the temperature of a location, the faster the wind speed will be and the higher the location, the faster the wind speed will be with the more dynamics wind movement [13][14][15].

Wind in architecture has the role of cooling space—wind motion functions as a heat spreader that determines the direction of the speed of the room. In addition, wind in architecture also functions as a medium that helps release heat in the human body. Therefore, an understanding of the determinants of aperture dimensions with a focus on wind speed is inevitable.

3.1. Wind Speed and Window Openings in Theory

The structure, lay-out, orientation and shape of a building affect wind speed in architecture—the building's structure in determining the wind functions as a receiver and limiter of wind currents. The lay-out and orientation of the building affect the formation of a positive-negative field in the wind area. The terrain determines the spread of the winds. The positive field is in the area of the coming wind, while the negative field is on the reverse side [16].

The building part and the building height are two elements in the tire structure. The building consists of two parts, namely walls and windows. The wall has two functions: deflect and direct the wind, while the window acts as a barrier and the entrance to the room [17]. The height of the building affects the movement of the wind that will enter the space. On two-thirds of the lower part of the building, the wind will pass through the side and through the building structure three times as fast with the direction of the wind that enters the room coming from above. On the upper third of the building, the wind flows to the top of the building with the direction of the wind entering the room from below [18].

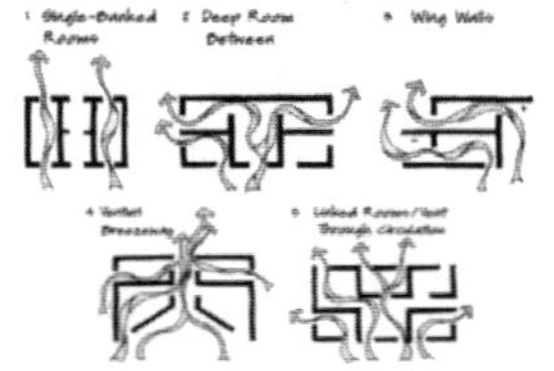

Figure 1. Walls as defenders of the wind

(source: Mark DeKay dan G.Z Brown. p: 229)

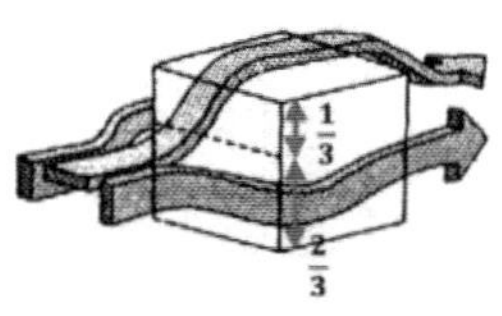

Figure 2. Illustration of wind movement against a building

(source: adaptasi dari *Terry S. Boutet*. p: 50)

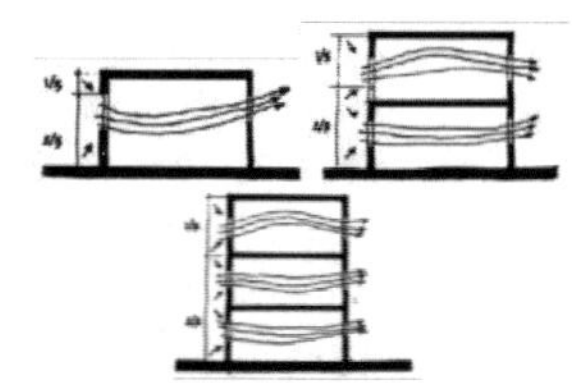

Figure 3. Illustration of wind movement into a building

(source: adaptasi dari *Terry S. Boutet*. p: 63)

The windows used in high-rise flat buildings are predominantly a projected sash window type. This type of window can open up to 100%. When the window is open at 100%, the direction of the incoming wind is the same as the direction of wind flowing outside the building; when the window is not open at 100%, the direction of the wind flow will follow the angle of the window opening. The determination of the size of the window or ventilation in theory is predominantly focused on finding the air rate, the inlet-outlet ratio, and the percentage of the wall area. This can be seen from several formulas or standards that are used globally or in Indonesia in particular.

In theory, there are at least two formulas based on the search for air velocity, inlet-outlet ratio, and the percentage of the wall area used in the calculation of window dimensions on the building façade, namely Terry S. Boutet's airflow rate search formula and B. Givoni's opening dimension formula, as follows[19][20][21]:

1. Terry S. Boutet : $Q = cfC_V Av$
   The search formula for Terry S. Boutet's airflow rate is focused on the search for air flow rate (Q), inlet area (Av), wind speed (V), and opening effectiveness (C_V) using the outlet-inlet ratio or C_V = 0.50 to 0.60 for winds perpendicular to the inlet or 0.25 to 0.35 for sloping winds with inlet, and conversion factors of 88.0 (cf).

IOP Conf. Series: Earth and Environmental Science **903** (2021) 012002 doi:10.1088/1755-1315/903/1/012002

2. B. Givoni : $\overline{V}_{(i)} = 0{,}45\,(1 - e^{-3{,}84x})V_{(o)}$

   The Givoni opening dimension search formula has a more complex calculation with a focus on airflow rate ($\overline{V}_{(i)}$), window to wall ratio (x), wind speed outside the building ($V_{(o)}$), and the basic natural logarithm = 2.72 (e).

Apart from these two formulas, there are also several standards or regulations governing openings in building façades in Indonesia, namely the Indonesian National Standard, as follows:

1. SNI 03-6572-2001: determination of the amount of ventilation based on the percentage of floor area and building class. A flat is a class-two building with a minimum opening size limit of 5% of the floor area of the room.

3.2. Rumah Susun Jatinegara Barat as the Research Object

Rumah susun buildings in Indonesia have at least some character and one of these is the window on the building's façade which has the same dimensions from the bottom to the top floor. The character of *rumah susun* is formed because *rumah susun* in Indonesia use a prototype system to speed up time and reduce cost of development. In addition, the type of windows of *rumah susun* predominantly has the same window shape, named projected sash window.

Figure 4. Projected sash window on bangunan *rumah susun* in Jakarta

Rumah susun Tambora (A), *rumah susun* KS Tubun (B), *rumah susun* Jatinegara Barat (C), *rumah susun* Pasar Rumput (D)

Rumah susun Jatinegara Barat have a characteristic location and environmental conditions that are common in high-rise flats in Jakarta, which are close to public transportation modes and located on a river bank. *Rumah susun* Jatinegara Barat is a 16-storey high and I-shaped mass building. 1st floor to 2nd floor are used as public areas and 3rd floor to 16th floor are used as residential areas with the building facing North and South following the river path.

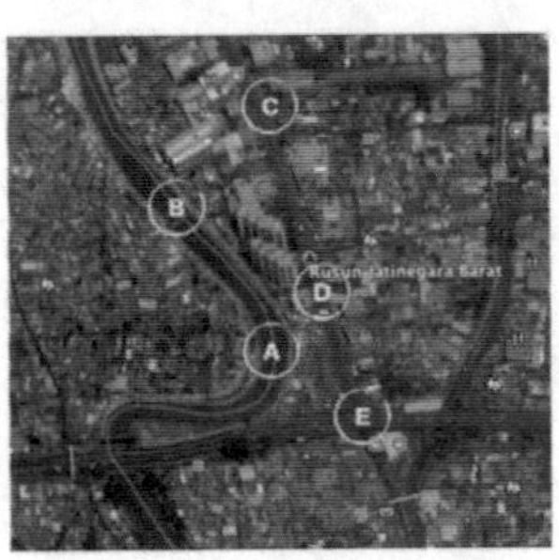

Figure 5. Location map of *rumah susun* Jatinegara Barat

Figure 6. Conditions around *rumah susun* Jatinegara Barat

The building of *Rumah Susun* Jatinegara Barat has several existing conditions: (1) the maximum opening is 10°; (2) most of the occupants close their windows due to the wind; (3) some of the windows are damaged due to strong winds; (4) the windows are single windows with the projected sash type; and

(5) each of the residential units has a size of 5.4 m x 5.95 m, a room height of 3.15 m with a window size of 1.4 m x 0.84 m on the main façade located in the two main bedrooms of 2.2 m x 2.6 m.

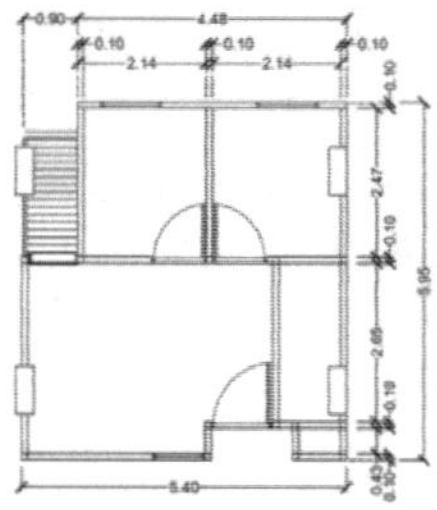

Figure 7. Residential unit floor plan of *rumah susun* Jatinegara Barat

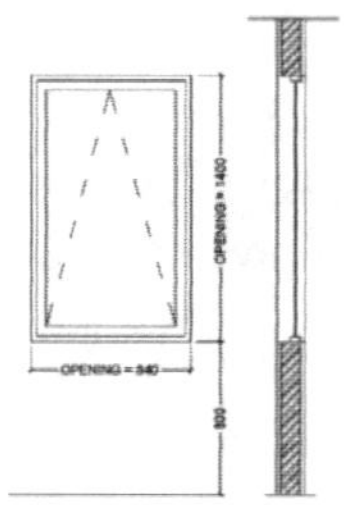

Figure 8. Windows detail on building façades

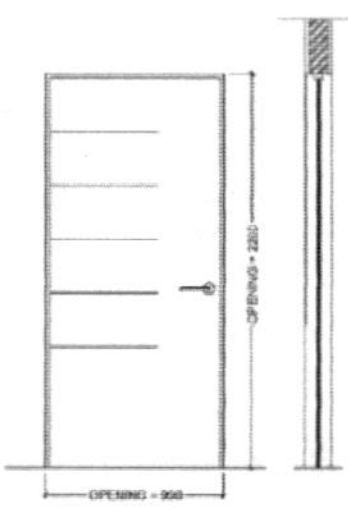

Figure 9. The main doors of the residential units

The configuration of the building mass of *rumah susun Jatinegara Barat* is I-shaped. It is forming a wind tunnel between the building towers. Field measurements carried out on 4th floor, 10th floor, and 15th floor with measurement times on 07.00 a.m. - 09.00 a.m., 11.00 a.m. - 01.00 p.m., and 03.00 p.m. - 05.00 p.m. show that the highest wind speed outside the building is 3.35 m/s (clear weather), 6.29 (strong wind), and 7.59 m/s (rain), as follows:

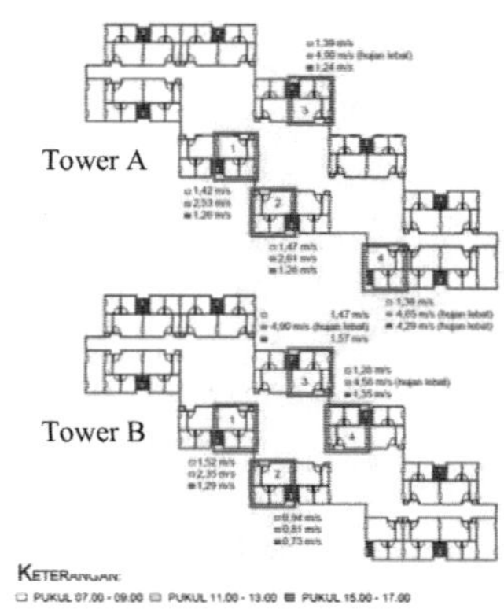

Figure 10. Wind measurement outside the building on the 4th floor

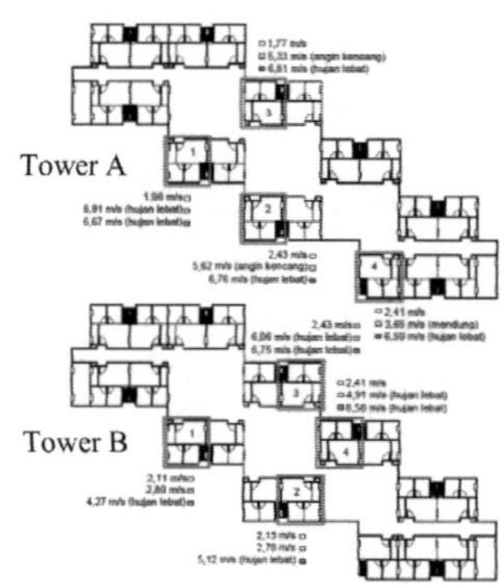

Figure 11. Wind measurement outside the building on the 10th floor

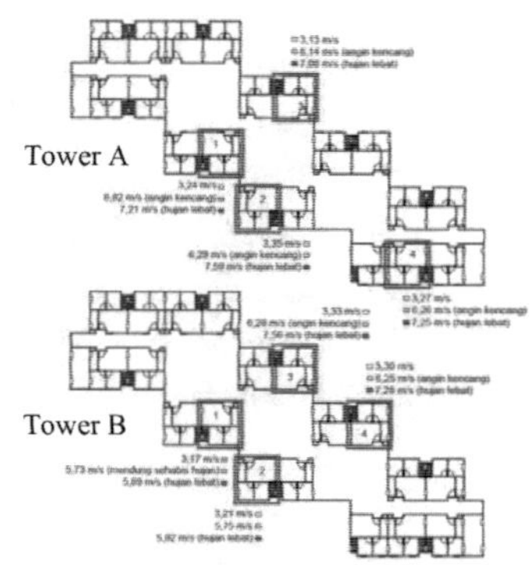

Figure 12. Wind measurement outside the building on the 15th floor

3.3. The Relationship between Wind Speed and Windows on the Façades of Rumah Susun Jatinegara Barat

Rumah susun Jatinegara Barat use a projected sash type with an opening area of 32,13 m^2 with a maximum angle of window opening is 10°. The openings located in residential units on the 3rd floor to the 16th floor. When there is and strong winds blow, none of the windows is used (closed), because the windows will fly with the wind and fall down if still used or opened.

Figure 13. Maximum angle of window - 10°

Figure 14. some of window opened when the weather is good.

Figure 15. Window closed when rains and wind speed intensity is high

Based on the theory briefly described above, it is found that high-rise buildings that consider wind speed in their planning must pay attention to four aspects, namely: the structure of the building (the wall as a directional tool for the wind path in the room), building lay-out, orientation of the building which determines the wind entry area, and the height of the building which affects the direction of the outside wind into the building. From these four elements, it can be derived that there are two main problems in *rumah susun* Jatinegara Barat, namely: (1) the lay-out of the adjacent buildings and the formation a wind tunnel, thereby accelerating the speed of the wind entering the wind tunnel area, (2) the orientation of the building side by side with the river, so that the winds that follow the river flow hit the façade of the building directly, and (3) monotonous openings indicating the height of the building are not the focus of planning. Both of these problems can be seen clearly through the calculation of the average wind speed through the calculation of the formula carried out on the 15th floor measurement with the assumption that the measurement results have the largest number when compared to the other three heights, as follows:

Tabel 1. Calculation of Openings through Formulas

Theory	Formulas	Result	Standard Boundaries
Boutet	$Q = cfC_VAv$ $Q = 88.0{,}35.0{,}52182.7{,}493737$	$Q = 120{,}439\ CFM$ **(does not meet standards)**	Min: 255,62 CFM Max: 780,02 CFM
	$Q = cfC_VAv$ $Q = 88.0{,}35.0{,}52182.14{,}07033$	$Q = 226{,}239\ CFM$ **(does not meet standards)**	Min: 255,62 CFM Max: 780,02 CFM
	$Q = cfC_VAv$ $Q = 88.0{,}35.0{,}52182.16{,}97835$	$Q = 272{,}876\ CFM$ **(meet standards)**	Min: 255,62 CFM Max: 780,02 CFM
Givoni	$\overline{V}_{(i)} = 0{,}45\,(1 - e^{-3{,}84x})V_{(o)}$ $\overline{V}_{(i)} = 0{,}45\left(1 - 2{,}72^{-3{,}84^2/3}\right)3{,}35$	$\overline{V}_{(i)} = 1{,}388\ m/s$ **(does not meet standards)**	Min: 0,2 m/s Max: 0,6 m/s
	$\overline{V}_{(i)} = 0{,}45\,(1 - e^{-3{,}84x})V_{(o)}$ $\overline{V}_{(i)} = 0{,}45\left(1 - 2{,}72^{-3{,}84^2/3}\right)6{,}29$	$\overline{V}_{(i)} = 2{,}606\ m/s$ **(does not meet standards)**	Min: 0,2 m/s Max: 0,6 m/s
	$\overline{V}_{(i)} = 0{,}45\,(1 - e^{-3{,}84x})V_{(o)}$ $\overline{V}_{(i)} = 0{,}45\left(1 - 2{,}72^{-3{,}84^2/3}\right)7{,}59$	$\overline{V}_{(i)} = 3{,}145\ m/s$ **(does not meet standards)**	Min: 0,2 m/s Max: 0,6 m/s
SNI	5% x room size	Bedroom = 2,352 $m2$ Drying room = 1,760 $m2$ Main door = 2,178 $m2$ Total opening area = 6,29 $m2$ **(meet standards)**	$5\%\ x\ 32{,}13 = 1{,}60\ m2$

Based on the calculation results, the total number of openings in one residential unit has met the minimum size. However, based on the formula calculation, the amount of entry opening in the façade has not been functioning optimally because with only 10° openings based on the calculation of the room area and the inlet: outlet ratio, the wind is not sucked into the room (the result of Q dan $\overline{V}_{(i)}$ *does not meet the standard; Boutet's formula shows that at an environmental wind speed*

IOP Conf. Series: Earth and Environmental Science **903** (2021) 012002 doi:10.1088/1755-1315/903/1/012002

of 7,59 the window opening can function optimally, but in this context it will be considered that the window still does not work because the environmental wind speed is far above the maximum threshold for a comfortable environmental wind speed for humans, which is 3.1 m /s), except all windows and doors in the residential unit are wide open.

3.4. Results

Based on the analysis carried out through theoretical studies of the relationship between high-level residential architecture and wind, case studies show that the determination of window dimensions on building façades can not only be seen through the area of space and the ratio of inlet: outlet openings in residential units, but must pay attention to the building's height as well. In addition, in theory, there is an affirmation that one of the elements influencing high-level architectural planning concerning the wind is the building's height. The functional shortcoming of windows on the façades *rumah susun* Jatinegara can also be seen from the temperature in the residential unit. The result of field measurements show that the average temperature of the residential unit is 30.75 °C. These results do not comply with the comfortable standard of room temperature which is in the range of 21 °C to 27 °C.

4. Conclusion

Changes in the shape of dwelling have a positive impact in the form of additional green areas and a negative impact in the form of increased environmental emissions. *Rumah susun* as one of the vertical residences in Indonesia is one of the buildings that has a negative impact on the environment. This occurs due to the use of prototypes that are not based on the environment and make the building performance become worse. In theory, understanding the relationship between high-rise architecture and wind can be seen through four aspects, namely: building structure (walls), lay-out, orientation, and the building's height. However, in the case of *rumah susun* generally, the building's height is not the focus of planning. This deficiency is also seen in the formulas and standards for determining the dimensions of the openings which only focus on three aspects, namely: air velocity, inlet:outlet ratio, and room size. Through the analysis of case studies, it can be seen that these three elements are not sufficient to determine the dimensions of the windows on the façades of high-rise flats, especially in Jakarta.

5. References

[1] World Meteorological Organization. 2008. *Climate Change 2007: Synthesis Report*. Switzerland: The Intergovernmental Panel on Climate Change. p. 5.

[2] World Meteorological Organization. 2015. *Climate Change 2014: Synthesis Report*. Switzerland: The Intergovernmental Panel on Climate Change. p. 88.

[3] Hoppit, Julian. 2011. *The Nation, The State, and The First Industrial Revolution*. Journal of British Studies. p. 307-331.

[4] Anthony, Wringley. E. 2018. "*Reconsidering the Industrial Revolution: England and Wales*" dalam Journal of Interdisciplinary Volume 49 Nomor 1, p. 9-42.

[5] Frick, Heinz; Suskiyatno, FX. Bambang. 2007. *Dasar-dasar Arsitektur Ekologis: Konsep Pembangunan Berkelanjutan dan Ramah Lingkungan*. Yogyakarta: Kanisius. p. 151-161.

[6] Frick, Heinz; Ardiyanto, Antonius, Darmawan, AMS. 2008. *Ilmu Fisika* Bangunan. Yogyakarta: Kanisius. p. 87-95.

[7] Kusumawardhani, Ratu Arum. 2012. *Sejarah Perumahan di Jakarta*. Jakarta: Dinas Perumahan dan Gedung Pemda DKI Jakarta. p. 3-4, 94-95.

[8] Rukmana, Dadang dkk. 2019. *Jejak Langkah Hunian Layak Indonesia* Jakarta: Direktorat Jenderal Penyediaan Perumahan Kementerian Pekerjaan Umum dan Perumahan Rakyat. p. 72.

[9] Alfata, Muhammad Nur Fajri dkk. 2020. *Pedoman Desain Pasif Rumah Susun: Untuk Iklim Tropis Lembap di Indonesia*. Jakarta: Kementerian Pekerjaan Umum dan Perumahan Rakyat. p. 7-13.

[10] Green Building Council Indonesia. Vol.1 Building Envelope. Jakarta: The Goverment of the Province of Jakarta Capital Special Territory. p. 3.

[11] International Finance Corporation (IFC). 2011. *Jakarta Building Energy Efficiency Baseline and Saving Potential: Sensitivity Analysis*.

IOP Conf. Series: Earth and Environmental Science **903** (2021) 012002 doi:10.1088/1755-1315/903/1/012002

[12] Brown, G. Z. 1994. *Matahari, Angin, dan Cahaya*. Bandung: Intermatra. p. 26
[13] Wirjohamidjojo, Soerjadi; Swarinoto, Yunus. *Iklim Kawasan Indonesia: Dari Aspek Dinamik-Sinoptik.* Jakarta: Badan Meteorologi Klimatologi dan Geofisika. p. 45.
[14] Boutet, Terry. S. 1987. *Controlling Air Movement: A Manual for Architects and Builders.* United States: R.R Donnelley &Sons. p. 41-138.
[15] Szokolay, Steven. V. 2004. *Introduction to Architectural Science: The Basis of Sustainable Design*. Oxford: Architectural Press. p. 74.
[16] Lechner, Norbert. 2015. *Heating, Cooling, Lighting: Sustainable Methods for Architects.* New Jersey: John Wiley&Sons. p. 293-295.
[17] Dekay, Mark; Brown,G. Z. 2014. *Sund, Wind, and Light: Architectureal Design Strategies*. New Jersey: Wiley. p. 229.
[18] Boutet, Terry. S. 1987. *Controlling Air Movement: A Manual for Architects and Builders.* United States: R.R Donnelley &Sons. p. 63.
[19] Indonesia. *Standar Nasional Indonesia: SNI 03-7013-2004 tentang Tata Cara Perancangan Sistem Ventilasi dan Pengkondisian Udara Pada Bangunan Gedung*. p. 3.
[20] Boutet, Terry. S. 1987. *Controlling Air Movement: A Manual for Architects and Builders.* United States: R.R Donnelley &Sons. p. 22.
[21] Givoni, B. 1976. *Man, Climate, and Architecture*. London: Applied Science. p. 22.

Acknowledgments
I wish to thank Prof. Gagoek Hardiman and DR. Rumiati Tobing, for guiding, encouraging, and convincing me to always write and believe in my thoughts. I hope that this article will prove useful for readers and can become the basis for one of the existing sciences, both in the academic world and for its practitioners.

Composing the noise space: Musical architecture of urban culture

F X T B Samodra[1] and I G Nugrahani[2]

[1]Department of Architecture, Institut Teknologi Sepuluh Nopember, Kampus ITS, Sukolilo, Surabaya 60111, Indonesia
[2]Balai P2P, Ministry for Public Works and Human Settlements, Gayung Kebonsari 50, Surabaya, Indonesia

fxteddybs@arch.its.ac.id

Abstract. In the recent climate change era of several tropical urban living and cultures, many slums are built on the railroad tracks accompanying the overheating environment. The people in that area can receive the noise radiation that was made from the railroad and the train wheels friction. The design research aims to manage noise with high intensity into the white noise, which was fit and valuable to increasing human life performance in daily life. It enhances through designing the attractive area at Malang urban area as the site and analyzing the environmental psychology approach according to adaptable barriers. This study uses audio editing and music notation software, Adobe Audition and Cubase, to capture the sound to support the design. The results highlighted that musical architecture could bridge the needs of the various people's living space and noise that cannot be moved and deleted. The play of materials, order, and composition of time that pay attention to the intensity and frequency of noise design project compose the rhythm of formal and spatial settlement design.

Keywords: architecture design, dynamic noise, railway, sound frequency, urban environment

1. Introduction

In the tropical urban environment, several settlements have been built along the railroad tracks and have become one of the destinations for the community to access creative industry services. The condition of the dwelling, which is on the side of the railroad tracks, needs special handling so that the building can respond to noise and accommodate user activities optimally. Noise waves are identical to disturbing sounds because they have different waves from human speech waves and musical sound waves. Noise is unwanted because it is unsuitable for time and space, affecting human comfort and health integrating with an urban heat environment [1]. It can affect human comfort and health due to three factors: frequency, consideration, and control. Noise depends on the sound source and the receiver which is described as urban activity and culture [2]. So it can be concluded that noise is a sound wave formed from various waves that do not go in rhythm so that noise can reduce the hearing intensity and interfere with the activities of the surrounding environment.

Noise on a moving train in the range of 85-100 dB depends on the train's speed. This noise comes from several things, including friction and vibration between the wheels of the train and the rail. The movement of the engine in the train and the concrete bearing vibration also have contributions [3]. These aspects produce internal noise emissions (in the train) and external noise or are exposed and accepted by the environment. In response to this, the Indonesian Ministry of Health divides several noise zones that are allowed to be in a city, namely zone A (35-45dB), which is designated for research, treatment, and hospital areas, Zone B (45-55 dB) for housing, education, and recreation. Zone C (50-70 dB) for offices, commerce, and markets, and Zone D (70-90) for industry, factories, railway stations, bus terminals, and the like [4].

In the context of the study being taken, Malang is selected for study location (Figure 1). In addition to the previous studies [1], [2], [5]–[7], the railroad area is located between the residential area where the noise threshold that humans can accept is 80 dB with an exposure of 8 hours per day, while 100 dB with time exposure to 15 minutes per day. It can be seen that the condition of the community on the site is very unhealthy because they get exposed to noise throughout the day. So that to deal with this, it is necessary to make a design that can bridge the site conditions without disturbing existing activities on the site. So that the invention is carried out so that the dwelling and shop areas along the site are not disturbed, but the building can still be responsive to noise and user comfort. Moreover, this research aims to respond to this noise; the musical architecture method is applied and used to translate, play, and regulate noise from the ground so that new experiences can occur and get activities that can run optimally on the site.

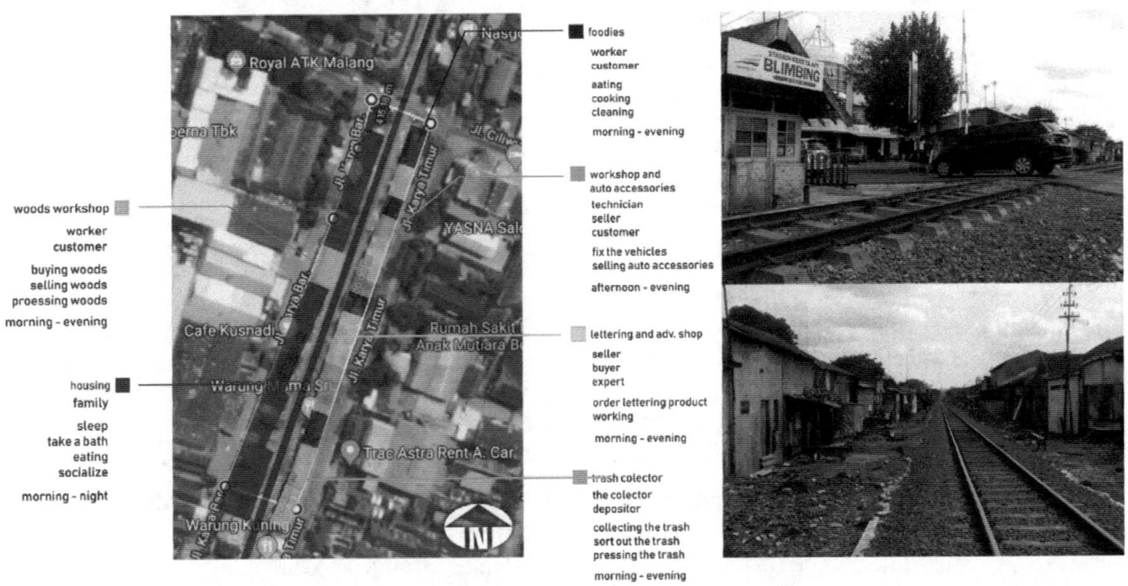

Figure 1. Condition of the site.

2. Methodology

This study conducts the site analysis to understand site conditions associated with noise and user activity. It is carried out by considering climate, culture, time, user activity, and movement of noise sources [8]. The method could be explained by research phases as follows:

1. *Site analysis,* it can be seen the influence of location and climate responding the climate change, zoning of buildings based on utility, zoning of buildings based on users, zoning of vehicle movements and noise sources, time-based noise points, and the intensity of noise exposure given by noise sources to site users. Intervention on the design result is adjusted to the synthesis of scores and the intensity of noise exposure, the function of the building, the users, and their needs. All are the illustration of urban culture.
2. *Field observation and measurement* use sound recording and sound level meter.

3. *Audio program and music note analysis are analyzed by* Adobe Audition and Cubase. It will be translated to the frequency analysis and site barrier control and architectural design based on the following domain to domain transfer analysis (Figure 2).

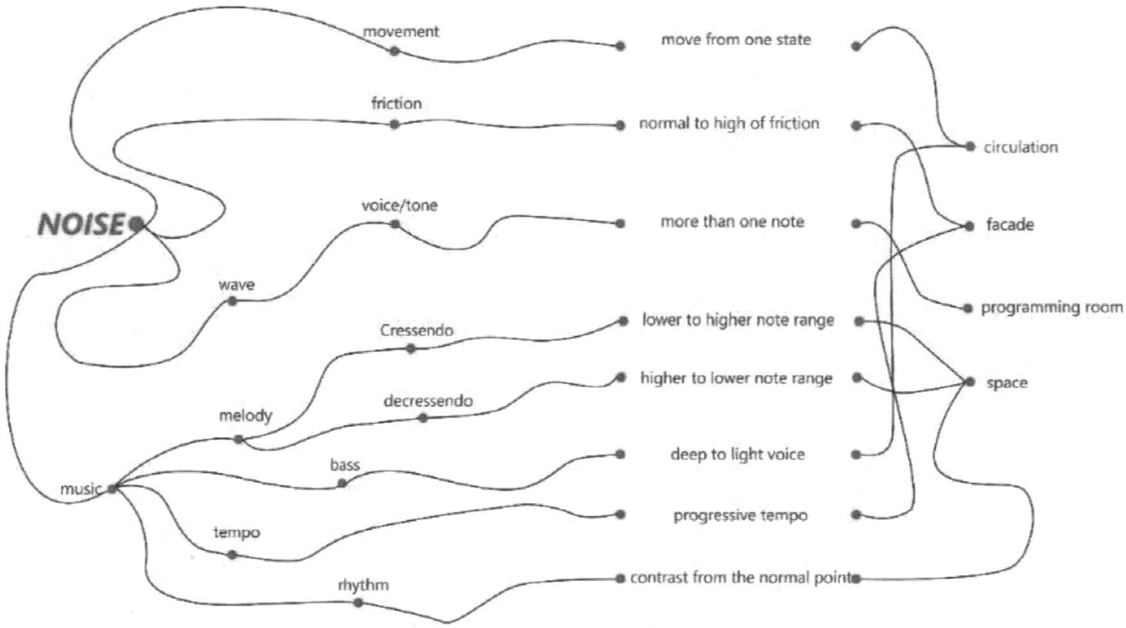

Figure 2. Domain to Domain Transfer Method.

3. Results and Discussion

3.1. Site analysis on environmental acoustics

The data inputted in the site analysis are area climate (radiation, temperature, rainfall, wind, topography, vegetation), site effects (site use and vehicles, user activity schedules, noise parameters, susceptible time to noise exposure, and building data. around, traffic, and density), schedule of train movement of pedestrian movements. In addition to the previous findings [9], the results that occur are synthesized so that they are able to illustrate what will happen in the design (Figure 3).

Music is a tone or sound that is arranged to contain rhythm, song, and harmony. It can be defined as the essential power which is very effective to calm and bring inspiration to many people. Through this definition, it can be understood that music is a group of sounds arranged so that it becomes a harmonious rhythm and can influence human emotions and performance. Music can also be interpreted as a part of human life and development. So that concerning the architectural science clusters, sound design can enrich the building experience and research how buildings and spaces can become an instrument.

There are similarities in the two groups of knowledge above, both of which discuss rhythm and harmony. On the architectural side, the emphasis is on the physical form of rhythm and harmony. At the same time, musicals pay attention to things that are connected to the user experience of the space. One method that can become a bridge for the two families is musical architecture. Musical architecture is a method that can unify space and sound into an enjoyable instrument and optimize the usability of space. There are several ways to apply to the musical architecture method. ONL's acoustic barrier uses a musical architecture method by combining a folding architecture to support the design implementation according to the design site conditions [10]. Le Corbusier's The Philips Pavillion uses aural-visual experience to apply the musical architecture method, to adapt it to the occupants' requirements and occupations.

Meanwhile, in applying the musical architecture method, Joel Sanders Architect uses a shared soundscape method in designing a Sound Lounge by taking into account the location and activities of the user. From the precedent results above, it can be found that the application of the musical architecture method is highly dependent on the conditions of the site, the users, and the activities that will be accommodated on the site (Figure 4). In addition to the previous finding [11], the user response to the environment procedures follows the attributes of environmental psychology; Noise volume, estimation of the occurrence, and the capability for controlling the source.

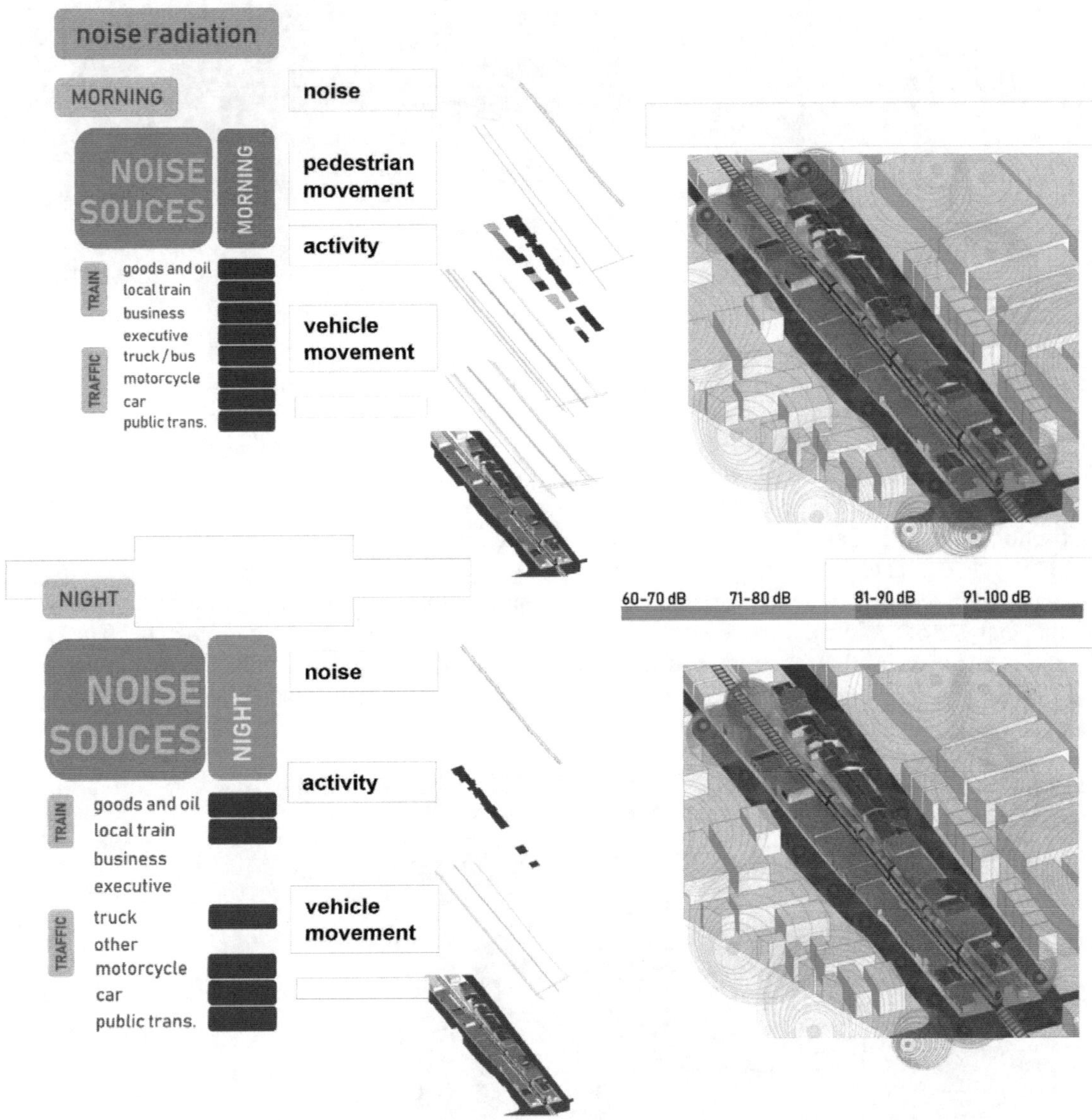

Figure 3. Noise radiation in daytime and nighttime.

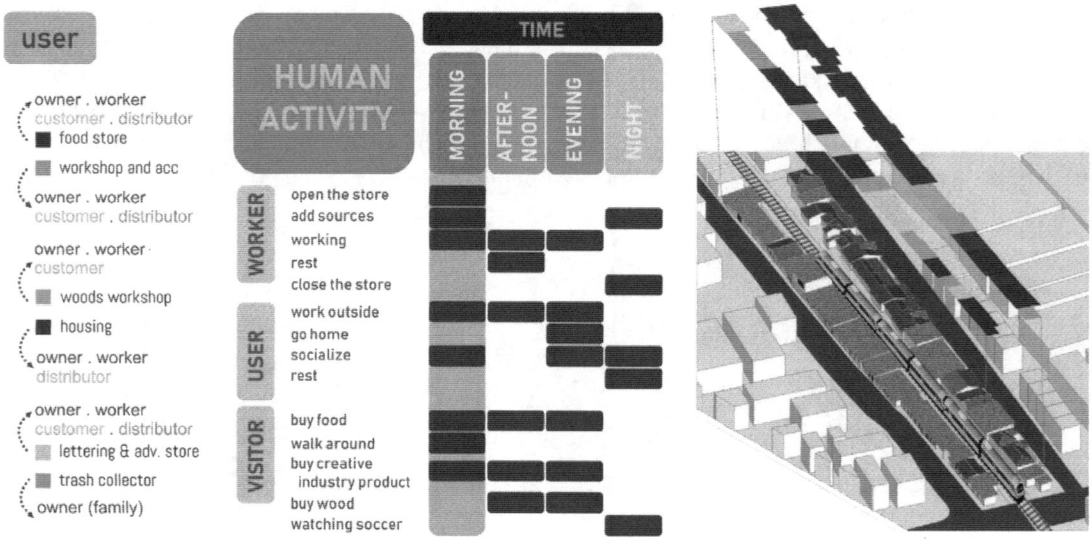

Figure 4. User analysis.

3.2. Experiment on acoustic simulation

On the shoulders of the site directly opposite the railroad tracks, initially, there are no significant restrictions. The site's safety against exposure to noise exposure, dust, and the movement of people and trains is not guaranteed (Figure 5). The design will apply a barrier capable of declining noise and increasing the security value for settlements. Barrier site scraping in response to noise is designed based on the analysis of scores and the level of intensity of noise exposure received on the site. The barrier's design also considers the function shaded on that part of the site so that the barrier along the site has various heights, and several sides expose the train's movement. The material applied to the barrier design is a gabion wall that contains tightly arranged boulders.

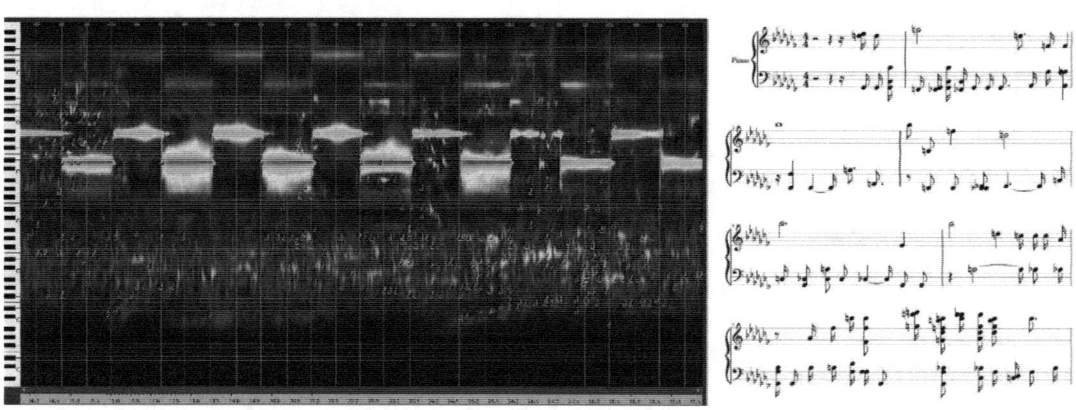

Figure 5. The recorded noise translation process for music notes and scores.

3.3. Dwelling concept translation from musical architecture

Reaching acoustical comfort is the same effort on integrating thermal in anticipating urban heat islands. The comfort shift will be predicted as a resilient idea [12]It is a channel for airflow in and out and

IOP Conf. Series: Earth and Environmental Science **903** (2021) 012003 doi:10.1088/1755-1315/903/1/012003

reduces incoming noise energy in responding to climate change (Figure 6-7). On the part of the site with relatively high noise emission intensity, the design is supported by a barrier in an acoustic glass. This is designed according to the direction of the noise. It is used to absorb most of the noise that passes into the building so that only a tiny portion of the noise can penetrate into the building. A custom rooster is also carried out to handle noise, especially from the building area facing the street, apart from thickening the walls.

The relationship of controlling noise for every sound frequency has consequences in a contradictive way for absorbing the wavelength as its dimension. For example, for sound attenuation of the most critical frequency, 1000 Hz, the building material must provide around 0.4 m of the hole as a noise trap. The lower frequency gives the bigger absorber or cavity on the building material or façade for reducing the noise. The music note is an illustration of tone, low or high frequency, and its intensity. Therefore, it is essential in architecture design which is translated into all the building elements, especially the frontal direction to the noise source.

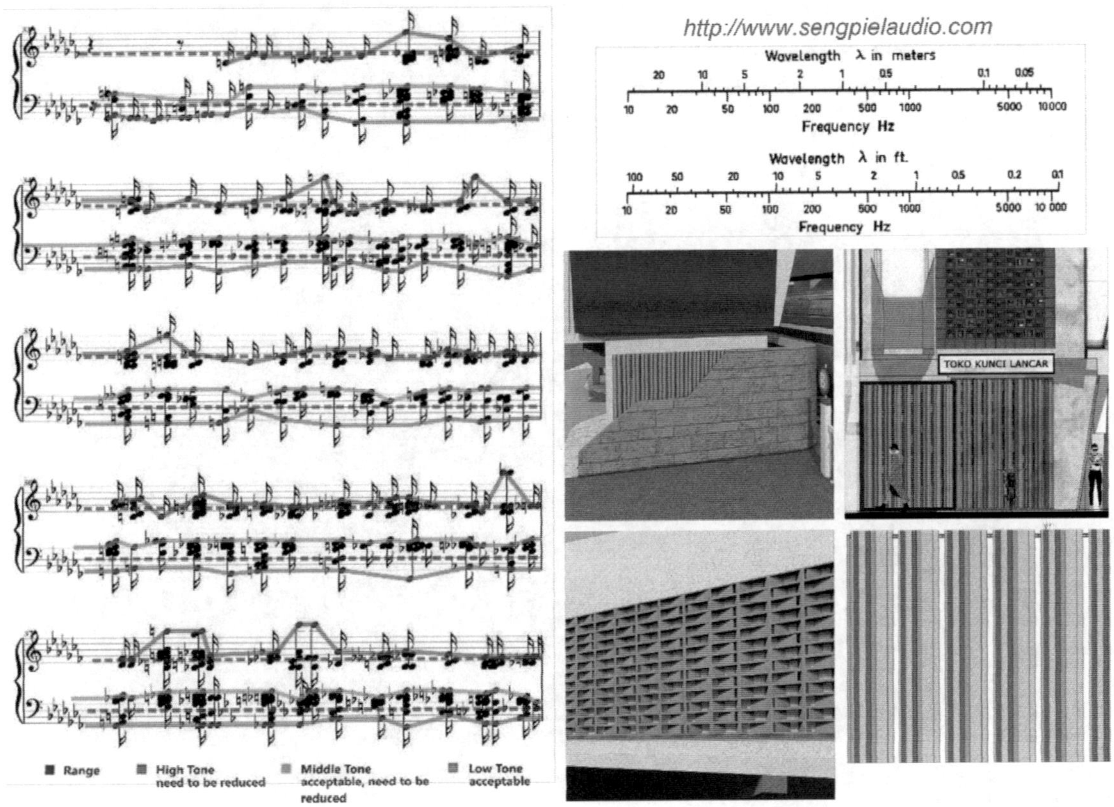

(a) Music score design (b) Architectural translation

Figure 6. Music score and architecture relationships.

The treatment of the dwelling roster design in the area is not jammed. The shape of the roster pattern depends on the point of road congestion. The roster is made of concrete. On the 1st floor, there is a vibrating panel to reduce noise emission energy that enters through vibrations. The concept of this vibratory panel adapts the idea of vehicle exhaust, where the gas boom that makes a loud sound is reflected at a certain angle and distance along the body of the exhaust until the sound that comes out is minimized. This panel is mainly placed in the area facing the highway, so that noise emissions from road congestion can be minimized and do not interfere with activities on the first floor with an open concept (Figure 8).

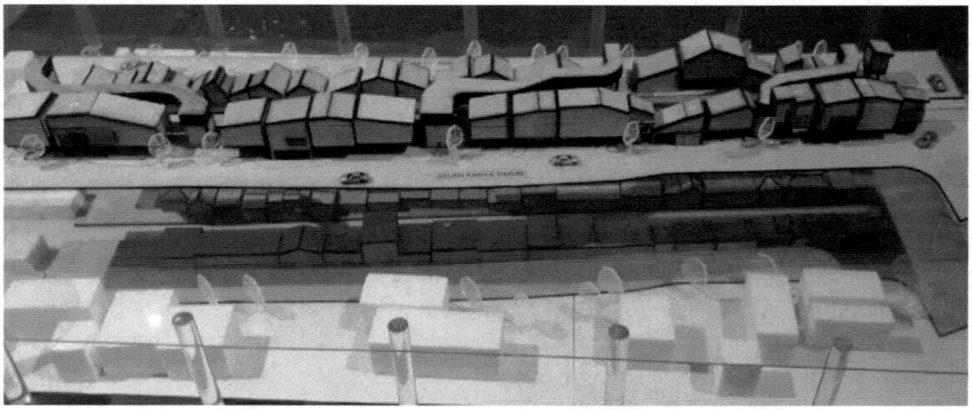

Figure 7. Existing and the proposed model.

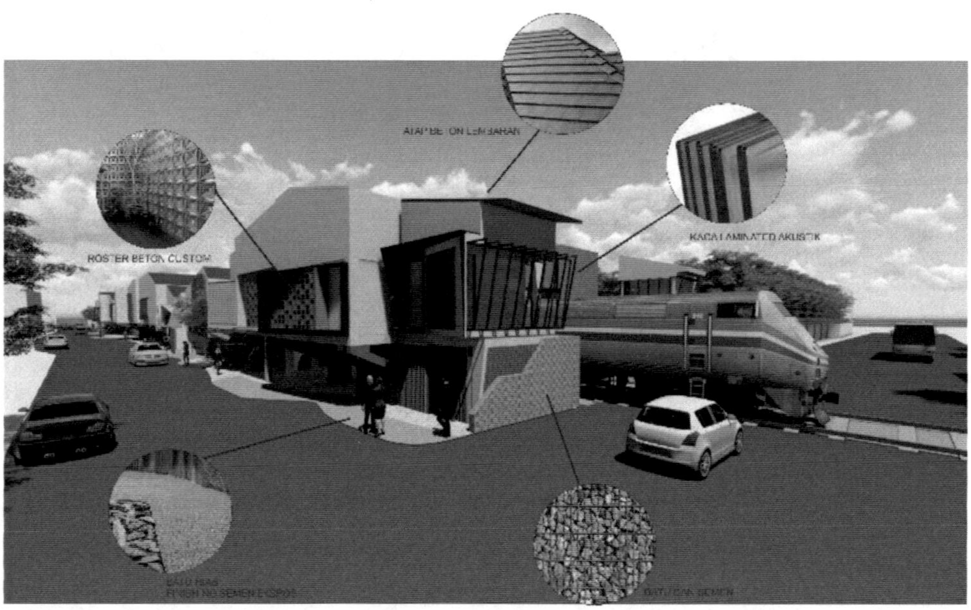

Figure 8. The existing and proposed model.

4. Conclusion

The musical architecture method is one method in identifying and changing the sound into an architectural masterpiece. In this design, the application of the musical architecture method is used to convert noise into architectural works by arranging and combining noise with other connected factors. The results of this method are syntheses that can be used as a reference in changing the design form. It lays out the design, makes patterns in the design, determines the design direction, and determines the materials and furniture in the design. So that the results of the building design can be responsive to noise emissions that occur, excess, and cannot be moved.

Acknowledgments

This research is part of Excellent Primary Research of Higher Education, Penelitian Dasar Unggulan Perguruan Tinggi No.933/PKS/ITS/2021. The authors gratefully acknowledge this financial and technical support.

References

[1] F. X. T. B. Samodra, "Analysis of resilient design by thermoacoustic adaptation of tropical urban model," *Journal of Architecture and Urbanism*, vol. 41, no. 4, pp. 305–315, 2017, doi: 10.3846/20297955.2017.1413960.

[2] J. Fallmann and S. Emeis, "How to bring urban and global climate studies together with urban planning and architecture?," *Developments in the Built Environment*, vol. 4, p. 100023, Nov. 2020, doi: 10.1016/j.dibe.2020.100023.

[3] F. X. Teddy Badai Samodra, "Soundscape elaboration from anthrophonic adaptation of community noise," 2018. doi: 10.1088/1755-1315/126/1/012034.

[4] Regulation of the Minister of Health of the Republic of Indonesia, *Industry Work Environment Health Standards and Requirements*. 2016.

[5] D. Botteldooren, L. Dekoninck, and D. Gillis, "The Influence of Traffic Noise on Appreciation of the Living Quality of a Neighborhood," *OPEN ACCESS Int. J. Environ. Res. Public Health*, vol. 8, pp. 777–798, 2011, doi: 10.3390/ijerph8030777.

[6] T. Münzel *et al.*, "Environmental stressors and cardio-metabolic disease: part I–epidemiologic evidence supporting a role for noise and air pollution and effects of mitigation strategies," *European Heart Journal*, vol. 38, no. 8, pp. 550–556, Feb. 2017, doi: 10.1093/eurheartj/ehw269.

[7] M. M. Zefreh and A. Torok, "Theoretical Comparison of the Effects of Different Traffic Conditions on Urban Road Traffic Noise," 2018, doi: 10.1155/2018/7949574.

[8] R. M. Rehan, "HBRC Journal The phonic identity of the city urban soundscape for sustainable spaces The phonic identity of the city urban soundscape for sustainable spaces," 2019, doi: 10.1016/j.hbrcj.2014.12.005.

[9] P. Tassi, O. Rohmer, A. Bonnefond, F. Margiocchi, F. Poisson, and S. Schimchowitsch, "Long term exposure to nocturnal railway noise produces chronic signs of cognitive deficits and diurnal sleepiness," *Journal of Environmental Psychology*, vol. 33, pp. 45–52, Mar. 2013, doi: 10.1016/j.jenvp.2012.10.003.

[10] P. Parthenios, S. Petrovski, N. Chatzopoulou, and K. Mania, "Reciprocal transformations between music and architecture as a real-time supporting mechanism in urban design," *International Journal of Architectural Computing*, vol. 14, no. 4, pp. 349–357, Sep. 2016, doi: 10.1177/1478077116670743.

[11] F. X. T. B. Samodra, Irvansyah, and C. Erwindi, "Investigation on environmental adaptation of traditional lifestyle in the tropical city housing," *AIP Conference Proceedings*, vol. 2230, no. 1, p. 040008, May 2020, doi: 10.1063/5.0002374.

[12] F. X. T. B. Samodra, Irvansyah, and C. Erwindi, "Standard review and update for tropical comfort shift of the built environment," 2018. doi: 10.1051/e3sconf/20186704013.

IOP Conf. Series: Earth and Environmental Science **903** (2021) 012004 doi:10.1088/1755-1315/903/1/012004

Sustainable Development of Taman Harmoni Tourism Area, Karangasem based on Local Wisdom's Value

N R P Salain[1] and N M M Mahastuti[1]

1. Architecture Department, Udayana University

Email: ratih_prajnyani@unud.ac.id

Abstract Sustainable development is raised as the theme of the article which is then analyzed based on the value of local wisdom with three main pillars, namely economic, environmental, and social aspects. *Manik Ring Cucupu,* Tattwa, Susila, Upacara, and Tri *Hita Karana* are the values of local wisdom used in organizing sustainable development in Taman Harmoni. The purpose of writing this article is to know the existence of economic, environmental, and social aspects that are the pillars of sustainable development in the Taman Harmoni tourist area based on the values of local wisdom. The method used is Qualitative Descriptive with data collection techniques through direct observation, interviews, and literature studies. The result of this writing is sustainable development in the economic aspect prioritizes the natural sustainability of the environment which is assumed to be a "container" as well as the value of local wisdom Manik *Ring Cucupu* which perceives the built environment with a natural environment such as a fetus in the womb of a mother; Sustainable development in environmental aspects is planned based on the value of local wisdom, namely *Tattwa* (philosophy), *Susila* (ethics), and *Upacara* (religious rituals); Sustainable development in environmental aspects, dividing the area of Taman Harmoni into three areas namely *Parahyangan, Pawongan,*and *Palemahan* in accordance with the teaching values of Tri Hita Karana.

Keywords: Sustainable Development, Tourist Areas, Value of Local Wisdom

1. Introduction
Sustainable development is a development that can meet the needs of the present without endangering future generations in meeting their needs [1]. The implementation of sustainable development faces challenges, namely how to improve welfare in using natural resources wisely, so that natural resources can be protected and the needs of future generations can still be met. Circumstances that often lead to failure occur when there is an imbalance between development in economic aspects, social aspects, and environmental aspects. Still based on the same source, states that the success of a sustainable development is characterized by the results of environmental impact analysis and public participation performance. Therefore, this article raises the theme of sustainable development in Taman Harmoni as a tourist area based on the values of local wisdom believed by the local community, namely the Bugbug indigenous village community, Karangasem.

Taman Harmoni Tourist Area is one of the tourist destinations that have developed since 2016. This area is managed by a body appointed by the agreement of the Bugbug indigenous village community, Karangasem, namely Bugbug Village Tourism Development Agency (BP2DAB). Before 2016, this area has become a tourist destination but only for some circles, this is due to difficulties in accessibility. This area is located between two beaches namely, Pasir Putih Beach and Pasih Kelod Beach. With such an interesting natural scenery, this area has the potential to be developed sustainably in order to still be enjoyed by future generations.

According to Yoeti in Helpiastuti [2] a tourist destination must have three factors of tourist activities concept, namely (1) *something to see;* (2) *something to do*; (3) *something to buy.* Before being managed by BP2DAB, this area only had the first concept of something to *see,* in theform of a view of two beaches and the beauty of the sunrise and sunset seen from Bukit Asah. Based on the value of local wisdom and the pillars of sustainable development, Taman Harmoni has now been able to develop the concept of something to *do* and something *to buy.*

Bali is an area with a variety of intangible cultures. One of them is the value of local wisdom that is believed by the local community. So, in producing tangible culture there is always the value of local wisdom. Similar to the sustainable development that occurs in Taman Harmoni, each pillar such as economic, social and environmental aspects imbued by the value of local wisdom, is the concept of Manik Ring Cucupu, Tattwa, Susila, Upacara, and Tri Hita Karana which serves as the basis in conducting sustainable development

1.1 Purpose and Benefits of Writing

The purpose of writing this article is to know the existence of economic, social, and environmental aspects that are the pillars of sustainable development in the Taman Harmoni tourist area based on the values of local wisdom. The value of local wisdom in question is Manik Ring Cucupu, Tattwa, Susila, Upacara, and Tri Hita Karana. While the benefits of this writing can be a reference for academic community about sustainable development based on the value of local wisdom.

1.2 Writing Method

The method used in this writing is a qualitative descriptive method because it utilizes qualitative data to describe the physical existence of Taman Harmoni tourist area, Karangasem in an effort to realize sustainable development based on the value of local wisdom. Qualitative data in question are data collected through direct observation techniques, interviews, and literature studies on tourist areas, Taman Harmoni Bali, and sustainable development.

1.3 Concept

In the concept will be described the definition and understanding of Sustainable Development, Tourist Areas, and The Value of Local Wisdom with the aim of similar understanding between the author and the reader.

1.3.1 Sustainable Development

Citing the concept of sustainable development outlined in Our Common Future or Brundtland Report [3] is:

> "Humanity has the ability to make development sustainable to ensure that it meets the needs of the present without compromising the ability of future generations to meet their own needs."

The meaning is that people in planning and implementing sustainable development must be able to ensure that the fulfillment of current needs does not sacrifice future generations in meeting their needs.

In sustainable development there are three main pillars, namely, economic growth, environmental sustainability, and social sustainability

Figure 1. Three Main Pillars of Sustainable Development
Source: http://sim.ciptakarya.pu.go.id/p2kh/knowledge/detail/pembangunan-berkelanjutan

From the diagram above, it can be seen that sustainable development is successful if economic growth does not suppress social pillars and dredge environmental pillars. So that future generations can still meet their needs. The three pillars must go hand in hand in a balanced manner.

Keiner in Pertiwi [4] expressed the understanding of Sustainable Development is to ensure dignified living conditions related to human rights by creating and maintaining a wide range or alternative access in planning lifestyles. The principle of justice between current and future generations should be taken into consideration in the use ofenvironmental, economic and socialresources. Comprehensive protection efforts on biodiversity and genetic diversity should also be considered.

Meanwhile, according to Law No. 32 of 2009 [5] on Environmental Protection and Management, Sustainable Development is a conscious and planned effort that combines environmental, social, and economic aspects into development strategies to ensure the integrity of the environment and the safety, capability, well-being, and quality of life of current and future generations.

1.3.2 Tourist Areas
Based on the Law of the Republic of Indonesia No. 10 of 2009 [6], what is meant by a tourist area is an area that has a certain area that is built and reserved for tourism activities. So, the tourist area built must adjust to the tourism activities. Meanwhile, tourism activities are generally obtained based on the tourist attraction owned by the area. The condition of an area can be used as a tourist area or tourist attraction is that there must be an attractive tourist attraction so that visitors are interested to come. The attraction in question can be natural (natural) or artificial. In Oka A. Yoeti [7], there are three things that should be considered in the development and management of tourist areas so that it succeeds in becoming a tourism industry, namely:

(1) Attraction

It is something that must be prepared in advance so that it can be seen, and enjoyed. In this case, dances, traditional folk-art singing, traditional ceremonies, and so on. Attractions can also be referred to as an attraction so that people want to come visit a tourist destination among them, such as:

a. Objects available and present in the universe (Natural Amenities): Climate, Soil and landscape forms, shrub forests, flora and fauna, health centers such as mineral springs, hot springs, and mud baths.

b. Man made supply that adapts historical, cultural, and religious elements. Such as: historical monuments and past architecture, museums, art galleries, traditional events, housesof worship, etc.

(2) Accesibility

Tourism activities are highly dependent on transportation and communication. So, it is necessary to consider what types of transportation can be used in reaching tourist destinations and how smooth the communication system is. In addition to transportation and communication there are also transportation infrastructures such as roads, bridges, stations, and airports. The function of infrastructure is to connect a place with another place. The existence of prasarana transportation greatly influences thepace of transportation. The better the condition of transportation infrastructure, the more optimal the speed of transportation used.

(3) Amenities

Amenities or facilities are one of the aspects that have a big influence on activities in the tourism industry. One of the facilities in question is hospitality accommodation. Because, it is not possible for a tourism industry to develop without having hospitality accommodation. Facilities are a supporting tool in the development of the tourism industry but it is very important. This is because tourists need a resting place in making a visit from one tourist attraction to the next located in one area. In addition to hotel accommodation there are other supporting facilities, namely: restaurants, clean water, communication, entertainment, and security.

Yoeti in Helpiastuti [2] mentions a tourist destination must have three factors concept of tourist activities namely (1) something to see; (2) something to do; (3) something to buy. With each explanation is:

(1) The area must have something called "something to see", meaning that there must be a tourist attraction or tourist attraction that is different from others and has a characteristic (something to see).

(2) The area should be available something called "something to do", meaning there should be recreational facilities that can make them feel at home for longer in the area (something to do).

(3) The area should be available something called "something to buy", meaning that in the place should be available facilities for shopping, especially handicrafts from the area, as well as other supporting facilities.

1.3.3 The Value of Local Wisdom

Local wisdom is all forms of wisdom based on the values of goodness that are trusted, applied and always maintained for a long period of time (hereditary) by a group of people in a certain environment or region where they live [8]. Then according to Alfian [9] local wisdom is a view of life and knowledge and as a strategy of life in the form of activities carried out by local communities in meeting their needs.

While local wisdom according to Law No. 32 of 2009 [5] on Environmental Protection and Management CHAPTER I Article 1 point 30 are noble values that apply in the community's living system to among others protect and manage the environment sustainably. So, from these definitions, conclusions that can be drawn about local wisdom are all forms of wisdom, outlook on life, knowledge, life strategy, and noble values that are trusted, applied and always maintained for a long period of time (hereditary) as a living system of localpeopleto meet the needs, protect and manage the environment sustainably.

Local wisdom has a strong attachment and relationship to the local / locus / local / area / nature. Local wisdom was born based on the community's response to the natural environment and did not take place instantly. Therefore, the character of local wisdom is said to be dynamic and sustainable.

Bali has a variety of local wisdom that has been passed down through generations because it is believed to provide protection and safety for the local community, namely Manik Ring Cucupu, Tattwa, Susila, Ceremony and Tri Hita Karana which will each be described.

(1) Manik Ring Cucupu

Manik means fetus while Cucupu means Uterus. So that Manik Ring Cucupu can be likened to a fetus living in the womb of its mother [10]. Which the fetus can only survive through the intake of healthy

IOP Conf. Series: Earth and Environmental Science **903** (2021) 012004 doi:10.1088/1755-1315/903/1/012004

nutrients from the mother. This is perceived in the relationship between the built environment and the natural environment. The built environment is likened to content and natural environment as a container.

(2) Tattwa, Susila, and Upacara

The three terms above are part of the three basic framework elements of Hinduism, each of which has its meaning as philosophy, ethics, and ceremony (religious ritual). The basic framework of Hinduism teaches its people to be able to serve the Creator starting from understanding philosophy / teachings of Hinduism, behaving like Hinduism and holding religious ceremonies as a form of gratitude.

(3) Tri Hita Karana

There are three causes of happiness/ harmony. The parts are the harmonious relationship between man and God (parahyangan), man with fellow human being (pawongan), and human relationship with the natural environment (palemahan) [10].

RESULTS AND DISCUSSION

Taman Harmoni tourist area is located in the tourist area of Bukit Asah, Bugbug Village, Karangasem. It takes about 2 hours from Denpasar to reach the highland area of Bugbug Village, Karangasem. Bukit Asah is only 4 km from Candidasa Karangasem beach. Therefore, Taman Harmoni has the potential to be developed into a tourist area considering the many accommodations and facilities in the Candidasa Karangasem.

Taman Harmoni tourism area is managed by bugbug Village Tourism Development Agency (BP2DAB) since the end of 2016 with its chairman, Jro Gede Arnawa until now. BP2DAB is appointed by Bugbug Indigenous Village as an effort of indigenous villages in improving welfare but the community can still participate in indigenous activities. In the past, bugbug indigenous villagers with productive age continued their education and sought jobs outside karangasem area so that they were often not present in traditional activities. Therefore, the village took the initiative to be able to createits own pekerjaanfield in order todo the work of the local community and even invite some local people to be entrepreneurial in the field of tourism.

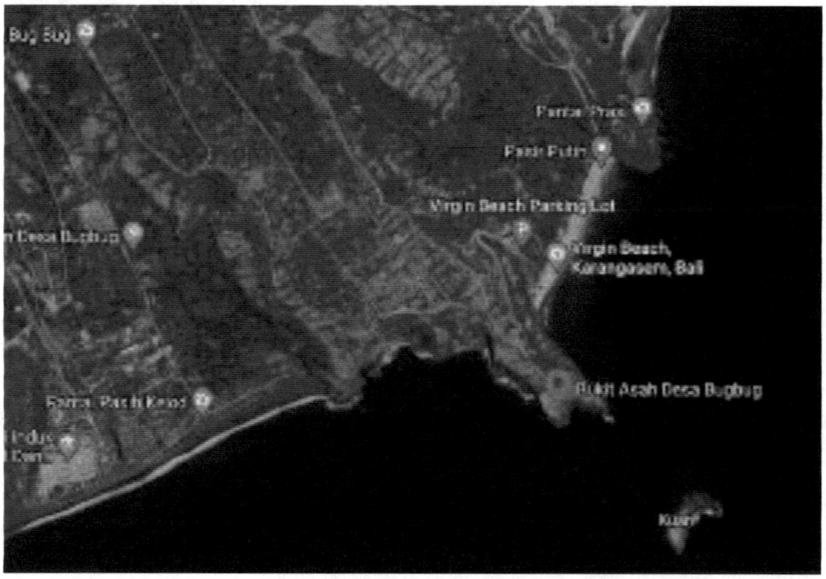

Figure 2. Satellite image of Bukit Asah area, Bugbug Village, Karangasem

source: https://www.google.com/maps/place/Bukit+Asah+Desa+Bugbug

In the satellite image can be seen Taman Harmoni is on the plateau of Bukit Asah Bugbug Village, Karangasem flanked by two beaches namely White Sand Beach and Pasih Kelod Beach with the presence of Kuan Island adds to the potential attraction of Taman Harmoni as a tourist area.

Before being managed by BP2DAB, the existing area of Taman Harmoni, is a tourist destination with natural attractions. Visitors or tourists who come can enjoy the view of White Sand Beach and Pasih Kelod Beach, because The Harmony Park is located in the middle. As the name implies, White Sand Beach has white sand, while for Pasih Kelod Beach has black sand. Besides being flanked by two beaches, from the Taman Harmoni area, visitors can also enjoy sunrise and sunset every day. This is what causes this area to be visited both morning and evening. Furthermore, the hot weather during the day and cold at night are also the next uniqueness for visitors. The existence of Kuan Island that looks like a whale from Bukit Asah is also one of the attractions for visitors who come. If in the past Taman Harmoni did not provide tents to stay, since the beginning of 2017, the manager provides tents with their contents, a set of tables and chairs in front of tents and campfires. Previously Taman Harmoni Bali did serve as a camping area, but did not have a name and tourists bring their own tents.

It is not wrong if the Bukit Asah area, especially in the Area of Taman Harmoni is developed into a tourist destination because this area has three things to become a tourism industry, namely:

1. Attraction
Often referred to as a tourist attraction because it is a contributing factor to the arrival of visitors or tourists. Based on the explanation, which includes attractions in Taman Harmoni Bali is: the existence of Taman Harmoni itself which is located between two beaches at once (Pasir Putih Beach and Pasih Kelod Beach) and is on the hill so that the visibility obtained is very wide towards the ocean. In addition, visitors can also enjoy the sunrise and sunset view from the top of the hill. Natural beauty that can not necessarily be obtained in other areas. Accommodation facilities are also widely available in Candidasa Beach area which is located approximately 4km from Taman Harmoni. So visitors who are more comfortable with accommodation with permanent character can choose in candidasa beach area.

Thus, the attraction contained in Taman Harmoni is everything that is available from nature to be managed and explored wisely by the manager. While the activities that can be done in this area in addition to camping rent a tent are: lighting a campfire, singing, playing guitar, recording pictures of yourself, fishing, riding a jukung boat, and grilling fish.

2. Accessibility
Since it is managed by BP2DAB, access to Bukit Asah is very easy to access. If in the past this area was often used as an offroad area because the terrain is steep, narrow and challenging. Now there is a road access to Bukit Asah with a width of 9 meters that can be reached by cars and motorcycles. With a distance of approximately 58.5 km through Jalan Ida Bagus Mantra, Denpasar for 2 hours visitors can already reach Taman Harmoni. The procurement of the road to Bukit Asah was initiated by BP2DAB at a cost of funding by the Bugbug indigenous villagers with the aim of increasing the number of visitors who come per day.

3. Amenities
The supporting facilities referred to in these amenities are hospitality accommodation (lodging), restaurants, clean water, communication, entertainment, and security. In Taman Harmoni, these three things (attraction, accessibility, and amenities) can be found in one area of this nature. Because, in this area is provided camping area for visitors who are saturated with the atmosphere of the hotel and challenged to enjoy the atmosphere of the outdoors. For restaurant facilities, in this area only provided small stalls of local residents Bugbug Indigenous Village, clean water and smooth communication, then for entertainment facilities BP2DAB plans to establish a cultural stage as a means or a community container in presenting traditional cultural arts activities, for security in taman Harmoni, employs local workers with a shift system.

IOP Conf. Series: Earth and Environmental Science **903** (2021) 012004 doi:10.1088/1755-1315/903/1/012004

Apart from attraction, accessibility, and amenities. Taman Harmoni as a tourist area can be seen from three factors of the concept of tourist activities, namely:

(1) Something to see

Everything that is classified on the attraction becomes an attraction to be seen by visitors in this Area of Harmony Park. Among them are Pasir Putih Beach, Pasih Kelod Beach, Kuan Island, the natural beauty of the environment and vegetation, as well as the view of sunrise and sunset that can be enjoyed from Bukit Asah.

(2) Something to do

Everything that visitors do when coming in this area is something to do. There are many things to do in this Taman Harmoni, namely: swimming at Pasir Putih Beach and Pasih Kelod Beach, fishing, staying in a tent, documenting pictures of yourself and view of the scenery.

(3) Something to buy

For the concept of something to buy in this area, there are no souvenir items that can be purchased but only drinks and snacks sold by small stalls. In addition, visitors can buy firewood as a means of support if they want to rent a tent and stay overnight.

From the three requirements of the tourism industry and the concept of tourism activities, it is very clear that taman Harmoni tourism area is worthy to be managed and developed in accordance with the principle of sustainable development based on the value of local wisdom.

Analysis

In the analyst is section, each problem formulationisoutlined to get the writing results. The problem of this writing is how the existence of economic, social, and environmental aspects that are pillars of sustainable development in the taman harmoni tourism area based on local wisdom values. Each aspect is analyzed based on the relevant value of local wisdom. In the economic aspect of Taman Harmoni is elaborated based on the value of local wisdom Manik Ring Cucupu, social aspects of Taman Harmoni are seen based on three basic frameworks of Hinduism, namely Tattwa, Susila, and Upacara. As for the environmental aspects of The Harmony Park is analyzed with the value of local wisdom from the concept of Tri Hita Karana consisting of Parahyangan, Pawongan, and Palemahan.

Sustainable Development on Economic Aspects based on The Concept of Manik Ring Cucupu in Taman Harmoni

One of the goals of development is economic growth that aims to prosper people's lives. Development and development carried out in the taman harmoni tourism area also has the same goal which is the welfare of the indigenous villagers of Bugbug, Karangasem. The traditional village wants to open new jobs and foster the creativity of the community in the field of tourism, in meeting the needs of its life. Considering the number of people with productive age wandering outside the area so that not all customary activities can be attended.

Therefore, the indigenous village appointed the Bugbug Indigenous Village Tourism Development Agency (BP2DAB) to manage. The development carried out by BP2DAB to date can be seen not only to reap profits but still heed the natural sustainability of the Bukit Asah area. Bugbug Indigenous Villagers and BP2DAB are convinced and well aware that Taman Harmoni tourist area is a tourist destination with its attractiveness in the form of natural potential so that the main thing to do is not to exploit the physical existence of the natural environment around.

Very relevant to the value of local wisdom believed by the bugbug indigenous villagers, namely Manik Ring Cucupu. In accordance with its meaning, Manik Ring Cucupu is the parable of a fetus in the womb of a mother who always needs nutrients from the mother in order to survive until birth. Similarly, the construction is carried out in Taman Harmoni. Taman Harmoni is assumed to be a fetus and the Bukit Asah area is assumed to be its mother. How can Taman Harmoni grow healthily if Bukit

Asah Area is not in a healthy and sustainable condition? That's the mindset of the Bugbug Indigenous Village community with regard to sustainable development in the Taman Harmoni tourist area.

Taman Harmoni is a built-in environment located in the Bukit Asah area as its natural environment. By maintaining the environmental sustainability of Bukit Asah, it can also indirectly improve the physical quality of "something to see" in taman harmoni tourism area which can ultimately increase economic growth because it attracts visitors to come.

The improvement made by BP2DAB since the end of 2016 has now had a positive impact on Bugbug Village, especially in economic development. According to the results ofan interview with the chairman of BP2DAB, Jro Gde Arnawa income in the economic sector is now experiencing a sharp surge. The number of visits that come every day has increased rapidly until before the pandemic. Especially for Taman Harmoni, tourists who come are getting more crowded on Saturday and Sunday. This is due to the easy access, adequate parking facilities, the rise of marketing on social media that can provide information to anyone, and the management of accommodation systems in Taman Harmoni, such as tent rentals and facilities in them .

Nominal revenue coming from Bukit Asah tourist area including in Taman Harmoni since 2017 is 140-200 million / month. This figure clearly shows an increase when compared to the previous revenue of 20 million / year (the result of an interview with Jro Gde Arnawa, August 2020).

From the data above, the revenue is divided for Bugbug Traditional Village by 20%, Taman Harmoni 60%, and to BP2DAB 20%. So that it can be stated that the arrangement and management by BP2DAB managed to make a positive contribution to the bugbug indigenous villagers on economic aspects without compromising environmental sustainability.

Sustainable Development on Social Aspects based on Tattwa, Susila, and Ceremony at Taman Harmoni

Sustainable development on the social aspects of Taman Harmoni tourism area is reflected in the culture, patterns, and activities of local people through generations in Bugbug Traditional Village, Karangasem. The three basic frameworks of Hinduism that serve as the basis of activity for his people. The three basic skeletons consist of Tattwa, Susila, and Upacara. Tattwa means philosophy / teaching / science believed by Hindus; Susila means ethics / behavior conducted based on Tattwa; and the last is the Upacara is a religious ritual offered as a form of gratitude.

For example, bugbug indigenous villagers believe in tattwa taught by Hinduism. One of them is the philosophy of Manik ring Cucupu. This philosophy has taught how we perceive the built/artificial environment as a fetus and the natural environment as the mother's womb. So that the arrangement and development carried out should lead to minimum intervention to the environment of the Taman Harmoni area. From Tattwa / philosophy believed by the bugbug indigenous villagers hereditary then created into a behavior / pattern of activity (Susila) from the local community.

The character of bugbug indigenous villagers has become accustomed to maintaining, and appreciating the existence of the surrounding environment. They always strive to preserve and maintain and utilize without intending to exploit their natural resources. This can be seen from the efforts of the village community in forming BP2DAB to manage the Bukit Asah area by considering the values of local wisdom as a tourist area whose profit is returned to the Indigenous Village.

They are trying to maintain the original contours of the Bukit Asah area along with the original vegetation. Road construction and parking lot arrangements also adjust existing land. The rental and establishment of tents in the Taman Harmoni area is also the implementation of Susila from bugbug indigenous villagers who seek to minimize the permanent establishment of buildings. So that the concept of lodging whose purpose is to enjoy nature naturally and blend with nature (not damage nature) can be realized here.

IOP Conf. Series: Earth and Environmental Science **903** (2021) 012004 doi:10.1088/1755-1315/903/1/012004

After the beliefs and patterns of activity that become a character of a society, there will be gratitude that wants to be offered and ask permission of salvation to the Creator who owns the universe. These offerings are often referred to as religious rituals or ceremonies. Bugbug Indigenous Villagers have various types of ritual ceremonies dedicated to Ida sang Hyang Widhi as a form of gratitude and hope to always be given safety. In the area of Taman Harmoni offerings performed (yadnya) consists of two types namely Nitya Yadnya and Naimitika Yadnya. For Nitya Yadnya is done daily by offering rice and dishes after cooking. Then for Naimitika Yadnya done in accordance with the calculation of sasih / month, calculation of wara, wuku calculation, or the occurrence of events or events that are considered necessary to be performed offerings

Ngusaba Manda ceremony is held once a year, the day before Nyepi Adat in Bali Aga Bugbug Village, Karangasem. This ceremony has been passed down through generations by previous generations to this day. The purpose of the Ngusaba Manda ceremony is to maintain the balance of the universe, namely, between Bhuana Agung and Bhuana Alit. The meaning of the word Manda is mepeileh; walking around the village boundary. The Teruna (teenage boys) and Daha (teenage girls) walked hand in hand around Bugbug Village starting from Pura Puseh in Hulu then to teben and back circling the Bale Agung Temple three times. After circling Bale Agung Temple three times, prayer ceremony began. After the prayer, the Teruna and Daha will return to each banjar to perform the Megibung procession. Because it is too vast, the procession around Bugbug Village is conducted at the boundary of traditional Balinese Aga settlements and focused on Bale Agung Temple. And the policy issued by the Traditional Village is when the Manda Ceremony is held tourist areas in Bukit Asah including Taman Harmoni should be closed to honor and appreciate the course of the ceremony. The policy was accepted sincerely by BP2DAB as the manager of Taman Harmoni. From this explanation, it can be understood that the manager and the people of Bugbug Indigenous Village strongly support the ongoing economic growth efforts but do not suppress the pattern of socio-cultural life that occurs

Thus,the positive impact felt by the people of Bugbug Karangasem Indigenous Village on development in the social aspect is a sense of confidence, togetherness, family and love for their homeland. People are also increasingly diligent and active in traditional activities in banjar because they do not have to wander outside of the village. Ceremony activities are also always held as a manifestation of gratitude for the grace they get.

Sustainable Development on Environmental Aspects based on Tri Hita Karana on Taman Harmoni

Tri Hita Karana is one of the local wisdom whose concepts and meanings are widely applied to every breath of Traditional Balinese Architecture. Tri Hita Karana is the three causes of happiness (harmony). The parts are Parahyangan which implies the meaning of harmonious relationship between man and His creator, Pawongan which means harmonious relationship between man and man, and the last is Palemahan, a harmonious relationship between man and the natural environment.

Based on the concept of Tri Hita Karana, the land use arrangement in bukit Asah area, Bugbug village is distinguished into three areas, namely Parahyangan, Pawongan, and Palemahan.

Parahyangan

It is the relationship between man and his Creator that is harmoniously established. In Taman Harmoni Bali, Parahyangan form that can be seen in the area of Bukit Asah, Bugbug Village, Karangasem, there is a arrangement in the area of Pasir Putih Temple so that people who will worship get easier and safer access. In the past, access to this temple was considered quite difficult because the road has not been paved and cannot be traversed by four-wheeled vehicles. Then, for now parking facilities have been provided along with public toilets.

Figure 3. Arrangement of Pasir Putih Temple as a form of Parahyangan in the area of Bukit Asah, Bugbug Village, Karangasem

source: https://www.google.com/maps/place/Bukit+Asah+Desa+Bugbug; Salain NRP, 2020

From the results of the arrangement, bugbug indigenous villagers feel the ease and comfort in performing ritual ceremonies and prayers at Pasir Putih Temple. The arrangement was implemented using funding from the Bugbug Indigenous Village community on the basis of initiatives from BP2DAB.

Pawongan

Pawongan definition is a harmonious relationship between human beings. For Pawongan area, it can be seen from community activities (social relations) conducted in an area. In the area of Bukit Asah, Bugbug Village, Karangasem, Pawongan area can be seen from

A. Ticketing area, due to social activity between guards and visitors
B. Small stalls providing food and drinks along the way to Pasir Putih Beach
C. Bali Harmony Park area including camping area
D. Pasir Putih Beach Area
E. Pasir Putih Temple Area

In the arrangement carried out by BP2DAB, Pawongan area is also prioritized for the convenience of visitors and local communities as well as the sustainability of the tourist area of Bukit Asah, Bugbug Village, Karangasem. In Taman Harmoni Bali, pawongan area can be distinguished by its function, namely, parking facilities, small businesses areas (stalls by local people), Taman Harmoni Bali office, camping area, and toilets.

IOP Conf. Series: Earth and Environmental Science **903** (2021) 012004 doi:10.1088/1755-1315/903/1/012004

Figure 4. Pawongan area arrangement on camping area in Taman Harmoni Bali
source: Anasthasia SAPC, 2021

Palemahan

It is a harmonious relationship between humans and the environment. Man can live his life by utilizing what has been produced by nature but, it would be nice if used wisely. Similarly, in the Bukit Asah area, Bugbug Village is managed as a tourist area with a natural view as an attraction, so that the sustainability of nature must be maintained and maintained if it wants to remain sustainable. The existence is very understood by the people of Bugbug Village and BP2DAB in organizing the Bukit Asah Area. The first thing they do is create and widen the road to facilitate access to achievements so that they can be traversed by four-wheeled vehicles. In the past, a narrow and winding road had to be traversed by visitors to reach Pasir Putih Beach. Not infrequently, Bukit Asah is used as a media as a plot of offroad events for adventurous communities. On the basis of a joint agreement between the community, village equipment, and the manager, the end of 2016 the Bukit Asah area began to be organized to improve the welfare.

The opening of the road is done at the cost of self-help from the bugbug indigenous villagers. The road planning does not use a team of experts, they only follow the existing road pattern (road pattern to Pasir Putih Temple). The road is widened to 9 meters and hardened with asphalt material. The gardens along the main road are made terraced according to the height of the ground level. The arrangement has actually changed the original form of Bukit Asah, but the changes that occur are expected to produce a positive impact on the community and the natural environment of Bukit Asah itself. They still strive to maintain the topographic shape by adjusting the original slope resulting in a terraced shape.

Figure 5. Main roads and landscape arrangements along the side of the main road

The authenticity and sustainability of Pasir Putih Beach and Bukit Asah is strived to remain sustainable until it can be enjoyed by the next generation. In addition to the utilization of the view as a tourist attraction, sustainability is also useful for the balance of natural ecosystems that have an impact on the surrounding community and future generations. Awareness of sustainable development seems to have been quite understood by the manager of Bukit Asah area and its surroundings, because the concept of Tri Hita Karana has always been used as a foundation in structuring and developing.

Figure 6. View Pasir Putih Beach, Bukit Asah area, Bugbug, Karangasem

Figure 7. Camping area laid out terraced by following the shape of the slope of the hill

In Taman Harmoni, the harmonious relationship between humans and the surrounding natural environment can be seen from the efforts of the manager in organizing camping areas that follow the shape of the slope of the hillside with a view leading to arah ocean. The terraced form also aims to optimize the viewpoint captured by visitors. So that all visitors staying in the camping area can enjoy the same view without differences.

Figure 8. Gazebo on the edge of Bukit Asah, at Taman Harmoni Bali

In addition, the form of harmony can be seen from the lack of permanent buildings of concrete built. The existence of permanent buildings in this area is found in public toilets and offices. Other buildings use materials of natural origin such as wood, bamboo and reeds. The trees and plants on this site also retain the ones that have existed since time before. As for the new one, the type of trees planted is a type of local vegetation from Karangasem.

Figure 9. Taman Harmoni Office Building and Toilets in Camping Area

Conclusions and Recommendations

The article with the title Sustainable Development of Taman Harmoni Tourism Area, Karangasem based on Local Wisdom's Value,was raised because it adapts the theme of writing namely Sustainable Development. The object of research was chosen because currently Bukit Asah Area, Bugbug Traditional Village is being visited by tourists both local and abroad. Especially in Taman Harmoni because since it is managed by BP2DAB based on local wisdom values,this camping area has progressed in economic, social, and environmental aspects.

Taman Harmoni in Bukit Asah, Bugbug, Karangasem area is analyzed to determine the existence of economic, environmental, and social aspects that are pillars of sustainable development based on local wisdom values. The values of Local Wisdom that are used as a concept are Manik Ring Cucupu, Tattwa, Susila, Ceremony, and Tri Hita Karana.

Economic Aspects

Manik Ring Cucupu is the parable of a fetus in the womb of a mother who always needs nutrients from the mother in order to survive until birth. Similarly, the construction is carried out in Taman Harmoni. Taman Harmoni is assumed to be a fetus and the Bukit Asah area is assumed to be its mother. Taman Harmoni is a built-in environment located in the Bukit Asah area as its natural environment. By maintaining the environmental sustainability of Bukit Asah, it can also indirectly improve the physical quality of "something to see" in taman harmoni tourism area which can ultimately increase economic growth because it attracts visitors to come.

Income that enters from the tourist area of Bukit Asah including in Taman Harmoni since 2017 is 140-200 million / month. This figure clearly shows an increase when compared to the previous revenue of 20 million / year (the result of an interview with Jro Gde Arnawa, August 2020).

From the data above, the revenue is handed over to Bugbug Traditional Village by 20%, Taman Harmoni 60%, and to BP2DAB 20%. So that it can be stated that the arrangement and management by BP2DAB managed to make a positive contribution to the bugbug indigenous villagers on economic aspects without compromising environmental sustainability.

Aspek Sosial

Tattwa, Susila, and Ceremony are part of the Three Basic Frameworks of Hinduism that are always used as guidelines in life. Bugbug Karangasem Indigenous Village people believe in the philosophy / teachings of Hinduism (tattwa) and then do it for generations to become a character (moral) and present it back to the Creator as a form of gratitude through religious rituals (ceremonies). By believing in the

IOP Conf. Series: Earth and Environmental Science **903** (2021) 012004 doi:10.1088/1755-1315/903/1/012004

concept of Manik Ring Cucupu, they then formed BP2DAB with the concept of enjoying and blending with nature (the concept of lodging with tents, minimal permanent buildings), as well as carrying out the ceremony of Nitya Karma, Naimitika Karma, and Ngusaba Manda as a form of gratitude to the Creator is a tangible manifestation of sustainable development in the social aspect based on tattwa, social, and ceremony.

So that the positive impact felt by the people of Bugbug Karangasem Indigenous Village on development in the social aspect is a sense of confidence, togetherness, kinship and love for their homeland. People are also increasingly diligent and active in traditional activities in banjar because they do not have to wander outside of the village. Ceremony activities are also always held as a manifestation of gratitude for the grace they get.

Environmental Aspects
Based on the concept of Tri Hita Karana, the land use arrangement in bukit Asah area, Bugbug village is distinguished into three areas, namely Parahyangan, Pawongan, and Palemahan.

Parahyangan; arrangement and management is focused on Pura Pasir Putih by providing parking facilities and public toilets. Repair and physical maintenance of the temple is also carried out periodically. Access to the achievement of Pasir Putih Temple in the form of a 9 m wide road is also facilitated by BP2DAB together with indigenous villagers.

Pawongan; The manager provides an area as a place to conduct social activities so that harmonious relationships are established between human beings, be it between visitors, managers, and daily workers on duty. Can be seen in the trade area (small stalls), the arrangement of camping area in Taman Harmoni Bali so as to create comfort for visitors who stay, the arrangement of Pasir Putih Beach naturally also gives the impression of relaxation for lovers of beach view.

Palemahan; harmony between people and nature is realized by the arrangement of tourist areas that prioritize environmental sustainability. In a sense, all developments made should be able to consider the positive and negative impacts that occur to the natural environment. In order to be able to be enjoyed by the next generation.

The efforts that have been made by the manager is to follow the hill slope transis for the arrangement of the landscape, maintain existing trees as barriers and shady, and replant local vegetation in order to grow well in accordance with the type of climate and soil.

Sustainable development on the economic aspect should consider development in the social and environmental aspects. Initial planning must be agreed on the extent to which development and development is implemented and the impact caused to the environment. In the Bukit Asah area, especially Taman Harmoni, the beauty and authenticity of nature is an attraction for tourists. As long as the sustainability of Bukit Asah area can be maintained as long as tourism activities can take place.

Based on the results of the writing and conclusions above, the recommendations that can be given to the Bukit Asah area, especially Taman Harmoni are:

- Continuing the understanding of the community about the importance of preserving nature so that it can be enjoyed by the next generation
- Adding signboards in the hill area that informs to maintain cleanliness, do not scratch trees or facilities in the Bukit Asah area, do not hunt wild animals such as birds, squirrels, etc.,
- Adding an appeal for visitors to use masks to be able to enter the bukit asah area for the sake of shared health.
- Provide and add washtafel and garbage cans both along the main road, in the area around the beach, parking facilities, and the Taman Harmoni area

- And increase the number of trees, especially for the landscape along the main road and parking facilities provided

Reference

[1] N. Sunarta and N. S. Arida, Pariwisata Berkelanjutan, Denpasar: Cakra Press, 2017.

[2] S. B. Helpiastuti, "Pengembangan Destinasi Pariwisata Kreatif Melalui Pasar Lumpur (Analisis Wacana Grand Opening "Pasar Lumpur" Kawasan Wisata Lumpur, Kecamatan Ledokombo, Kabupaten Jember," *Journal of Tourism and Creativity. Universitas Jember,* vol. II, no. 1, 2018.

[3] W. C. o. E. a. D. (WCED), Our Common Future,, Oxford.: Oxford University Press, , 1987.

[4] N. Pertiwi, Implementasi Sustainable Development di Indonesia, Bandung: Pustaka Ramadhan, 2017.

[5] *Undang-Undang Republik Indonesia No. 32 Tahun 2009 tentang Perlindungan dan Pengelolaan Lingkungan Hidup.*

[6] *Undang-Undang Republik Indonesia No 10 tahun 2009.*

[7] O. A Yoeti, " Perencanaan dan Pengembangan Pariwisata," Jakarta, 1997.

[8] R. Njatrijani, "Kearifan Lokal Dalam Perspektif Budaya Kota Semarang," *Gema Keadilan,* vol. V, no. 1, 2018.

[9] M. Alfian, "Potensi Kearifan Lokal Dalam Pembentukan Jati Diri dan Karakter Bangsa," in *Prosiding The 5 th International Conference on Indonesia Studies:Ethnicity and lobalization*, Jakarta, 2013.

[10] N. K. A. Dwijendra, "Perumahan Dan Permukiman Tradisional Bali," *Jurnal Permukiman "Natah",* vol. I, no. 1, 2003.

[11] Y. Kristian, "Pengelolaan Objek Wisata Oleh Dinas Pariwisata Kabupaten Kutai Barat Di Danau Aco Kampung Linggang Melapeh Kecamatan Linggang Bigung," *EJournal Administrasi Negara. Universitas Mulawarman,* vol. V, no. 1, 2017.

[12] S. N. Pendit, Ilmu Pariwisata Sebagai Sebuah Pengantar Perdana., Jakarta: Pradnya Paramitha, 1994.

[13] A. Rapoport, Culture, Architecture, and Design., Chicago: Locke Science Publising Company, Inc., 2005.

[14] G. Suartika, J. Zerby and A. Cuthbert, Doors of Perception to Space–Time- Meaning: Ideology, Religion, and Aesthetics in Balinese Development. Space and Culture, Sage Publishing, 2018.

[15] I. N. Susanta and I. W. Wiryawan, "Konsep Dan Makna Arsitektur Tradisional Bali Dan Aplikasinya Dalam Arsitektur Bali," *Workshk Arsitektur Etnik dan Aplikasinya Dalam Arsitektur Kekinian,* 19 April 2016.

Commoning: Towards the Eviction of Kampung Bidaracina, Jakarta

Muhammad Luthfi Fikri[1] and Herlily[2]

[1] Department of Architecture, Faculty of Engineering, Universitas Indonesia, Depok 16425, Indonesia

[2] Department of Architecture, Faculty of Engineering, Universitas Indonesia, Depok 16425, Indonesia

Email: mluthfifikri@gmail.com

Abstract. Over the past decades, flood control has been providing the rationale for evictions in Jakarta. Underlying this study is a set of notions where the eviction plan would bring up uncertainty to their commons: the risk of possible destructions of their commons, the uncertainty of livelihood, or the dubiousness of securing a decent amount of compensation. Considering that, this study is trying to identify the scope of commons and commoning in long-envisioned yet unimplemented eviction of RW 04 Bidaracina, Jakarta, as well as to understand its relation to resilience in several forms. We conducted interviews to understand the aspirations of the residents of RW 04 Bidaracina to answer the questions of this research, which concern mainly on how the notions of commoning are manifested in a community expecting a rather uncertain eviction. The result corresponds favourably with existing studies on kampung as urban commons.

Keywords: commons, commoning, eviction, informal urban settlement, kampung

Introduction

1.1. Current Conception of Commons

In general, commons are known as public resources that are used and exploited by groups of people or communities where the commons themselves do not object to the formal sector (government or other formal institutions) and are not capital commodities [1][2], although historically commons refers to pastoral or grazing land for cattle which is used and shared by a community of herders [3]. In 1968, Hardin theorized the poplular 'Tragedy of the commons': a situation in which users, who have free open access to a resource not governed by shared social structures or rules that govern access, act independently according to their self-interest and cause depletion of the resource, hence destroying the commons and communities that depend on it. Hardin theorized that in order to govern a commons, the community which relies to the commons have to be regulated [3].

The counter idea was proposed by Ostrom [4], proving the importance of the commons while investigating an answer for the well-known theory of "Tragedy of the Commons" by Hardin. Based on Hardin's dilemma, Ostrom offers 8 principles for how commons can be governed sustainably and equitably in a community, how communities co-operate to share resource use, the public goods and the future of the community [4]. Although the proposed eight principles of governing a commons is

IOP Conf. Series: Earth and Environmental Science **903** (2021) 012005 doi:10.1088/1755-1315/903/1/012005

considered to be a successful answer to Hardin's dilemma, the 'dilemma' (i.e. the concept of commons) itself expanded.

In recent years, commons is not only defined as shared physical resource. Harvey has a radical approach that defines commons as a resource that includes not only land or places where it is shared, but also language, social practices and construct, and other intangible social fabric [2]. According to Harvey, commons are fostered continuously as a relationship between certain social groups and environmental aspects (in the form of social or physical aspects) which are considered important for the lives and livelihoods of its users, the commoners [2].

Although the sustainability of commons happened to be essential as being the livelihood strategies for the socio-economic security of informal communities, the term commons has been broadly conceptualized to define multiple extents. The discussion of the idea of commons and commoning has become a broader discussion in the last decade as a criticism of urban planning, where urban planning is heavily inclined towards values of capitalism [2]. However, the understanding of commons and commoning concerning the evictions of riverside kampungs in Jakarta have not been thoroughly assessed, despite the importance of commoning as the strategy of livelihood for the urban informal community: what Simone (2014) conceptualized as "the urban majority" in Jakarta [5].

1.2. Plan for Bidaracina Eviction: A Circumstance to Understand Commons

To define and understand the concept of commons in the context of informal urban settlements in Jakarta, we argued that eviction, an event which could result in eradication of commons, could give us an idea of how informal settlers are thinking about their commons and its relation to their livelihood. As argued by Hellman, informal settlers, many of those who benefit from the commons, are in fact often unaware of the values of the constructed social capital as their own commons: they took it for granted [6]. However, when their commons are threatened by eviction plans, the communities which heavily relied in social capital created and nurtured in kampungs show reactions in different ways. On the one hand, uncertainty of eviction has resulted in residents in several areas that were about to be evicted to delay repairs to the damage that occurred in their homes. In a different case, the uncertainty has not discouraged residents from investing in a private property. On the contrary, citizens will instead make either physical or non-physical improvements or investments. Based on observations made by Hellman, private investments locals make in their dwellings and properties are partly an effort for them to be incorporated into society and to build a sustainable livelihood. In the latter case, commoning might also take shape in the form of ideas and connections to authority, to show gestures of resilience that citizens have. This has been observed by Hellman 2018, Padawangi 2019, and Dovey 2019, which argued that there were more efforts shown by the informal community who were threatened with eviction [6][7][8].

Looking at those instances where communities took actions when their commons are threatened, we saw the eviction plan of RW 04 Bidarcina, an informal neighborhood located on the eastern bank of the Ciliwung river, as an opportunity to recreate the observations. RW 04 Bidaracina and other informal settlements (*kampungs*) share a common "fate", that they were labelled as and stigmatized as the cause for flooding.

The underlying force behind Bidaracina eviction, as also was for other evictions, is that of flood mitigation. RW 04 Bidaracina is a kampung located on the eastern bank of Ciliwung river, which has long been haunted with eviction threat with the plan to construct water diversion project to divert excess water flow of Ciliwung river to eastern flood canal (BKT) [7]. The land acquired from the planned eviction is allocated for the construction of the Ciliwung- Easter Flood Canal (BKT) water diversion tunnel, which was planned to be able to divert the flood discharge of the Ciliwung River to BKT has been on the board since 2008 by the Center for the Ciliwung-Cisadane River Basin (BBWSCC) [9]. This development is a way to supposedly protect Jakarta from seasonal flooding (Ciliwung River especially the Cawang – Manggarai Water Gate) by reducing the flow of water during heavy rain to the nearby BKT canal. The diversion tunnel was estimated to evict families in RW 04 and RW 14 Bidaracina.

Research Methodology

With the ever-expanding scope of commons, the first question to be researched is the scope and limitations of the concept of commons and commoning from various literature studies in riverside kampungs, especially in the particular context where the commons themselves are under the threat of eradication. The second objective is to further trace and identify the commons and commoning processes that exist in RW 04 Bidaracina and how they might affect the built environment of Kampung RW 04 Bidaracina, especially the ones related to flood control and how commons could be utilized as a response to urban flooding; utilizing scope of commons gained from the literature review.

The data used in this study are primary data from distance interviews and secondary data from academic literature. We conducted distant interviews with two residents of RW 04 Bidaracina. Secondary data were obtained from the narrative review semi-systematic literature review (SSLR) method, which is a research method designed for topics that were conceptualized and studied by various groups of researchers in various disciplines and which made systematics impossible to do effectively [10]. Because it covers a wide range of topics and different types of studies, the research must be transparent and have a research strategy developed that allows the reader to judge whether the arguments to judge are reasonable, both for the topic chosen and from a methodological perspective [10].

We conducted interviews to two residents of RW 04 who played active roles in contributing to saving their commons from eviction: Pak Edi as secretary of the RW (neighborhood) and Mba Aci as part of the community's legal team. From the interviews with the two residents, we identified commoning in RW 04 and how the commoning process had an impact on the built environment.

Commons and Commoning: Scope and Boundaries in the Context of Kampung

This section discussed literature review regarding scope and boundaries of the concept of commons and the process of commoning in the context of informal urban setting in Jakarta, especially when they are threatened by eviction in the name of flood control. This section explores the notion of commons which not only includes a set of temporary resources - such as clean water, sanitation – but also access to political structures and private assistance and socio-economic security - that together provide sustainable livelihoods [11]. This section also explores the scenario of how commons might be destroyed and how commons and the practice of commoning might hold a significant role in reshaping the urban informal settings of Jakarta, especially in relation to urban flooding.

1.3. Commons and Urban Resilience

In the context of kampung as informal settlement, the identification of commons and commoning can be understood as the resilience shown by the residents. Resilience is defined by Jones as the ability to adapt to change, including its vulnerabilities [12]. Jones' definition centers on the notions of robustness, mitigation, and adjustment. The idea of commoning, in this sense, coincides with the idea of guerilla urbanism, where the sense of commoning also extends to the concept of reoccupying certain urban spaces as a tool for experimenting with various ways to produce a public, creating a mechanism to build inclusiveness for city residents [5]. The appropriation to reoccupy particular space in a city and make changes without the consent of formal administration is analogous with the acts of commoning of urban space.

The residents' efforts to make improvements to their personal and communal properties are a part of the struggle to show their resilience. This corresponds to the studies of Hellman (2018) and Betteridge & Webb (2019), concerning commoning and urban resilience respectively. In this case, residents of Kampung Pulo reflected the importance of a decent property as one of the most reliable evidence that the people living in them were not squatters, in other words, a resilient community. This community effort is a way of making their kampung as a whole to be remarked as a "well-established kampung", or "a hereditary kampung" because this kind of resilience is very much needed in the process of developing

IOP Conf. Series: Earth and Environmental Science **903** (2021) 012005 doi:10.1088/1755-1315/903/1/012005

their resources [6]. Establishing well-built private homes and kampung's facilities also paved the way for various forms of collective visions of ownership and of belonging to the community of their kampung. Despite this, many of those who benefit from the commons are unaware of their values [6].

Kampung improvement for the sake of community resilience may also leads to partial self-destruction by its very own residents, in this case the residents of kampung Tongkol. In the case of Tongkol, the resilience of the community in the form of security of land and property ownership are the result of the community's own attempts to change the image of their kampung from being 'slum' to 'clean and tidy' [7]. Unlike Kampung Pulo, Kampung Tongkol has a more spatial self-help approach to achieving resilience. The resilience shown by the residents of Tongol Kampung did not have any significant obstructions because the residents even voluntarily evict themselves. Dovey (2019) emphasizes the self-help willingness of the Tongkol community to discover other ways of 'playing politics' which allow results in the form of durability and resilience, even by demolishing and rebuilding the kampung partially [7]. Instead of simply struggling toward the authority of law through the legal advocacy process, 'playing politics' means that they have gained resilience through citizen-based alliances that the residents of Kampung Pulo do not possess.

Social and economic networks are shared resources that are jointly built and utilized by residents. What makes public awareness in forming a sense of community lies in the socio-economic network in the form of family, neighbors, social and religious gatherings, and community groups especially communities based on regional origin from various regions across Indonesia [14]. Apart from the social structure of the social safety net that is considered reliable, the location of the kampung also accounts for livelihood security. This argument is supported in the research of Sari et al. (2018) that 44% of people content to live in a kampung in Jakarta, namely Kampung Pulo, because residents always help each other [15]. Those who are content to live in Kampung Pulo for their strategic location make 42% of the residents, only two percent below the latter opinion [15]. It can be assumed that the socio-cultural and economic aspects do not overshadow each other. This is the embodiment of the notion of near-south, that commoning in Kampung Pulo does not only involve the use and development of shared resources (in this case the social safety net) but also provides a good place for the competition to emerge, where residents the kampung must be different from other residents in competing for a living. The houses of the residents of Kampung Pulo, part of which are used as places of business, are commons with 'a hint of competition' as a way for the residents to gain sustenance.

1.4. Assessing Right to The City: Legal Actions to Protect a Commons

In the context of evictions, the struggles of community representatives to file a class-action lawsuit to try to revoke the eviction show the resistance efforts made by them. This kind of resistance shows the awareness of the right to the city and ownership, but in the end, the residents will show another rhetoric: they want to return to their old ways of living and living in a kampung, to live as neighbors, to work together, and to restore the social networks they had previously built in their kampung [13]. Hellman (2018) translates this situation as residents underestimate the presence of neighbors and what is produced from being in a community of neighborhoods as things that are only taken for granted [6]. Solely when the commons and means of commoning were lost, residents could perceive the value of the importance of the social networks they built in their kampung as commons.

Class action lawsuits happened in various occasions, in Kampung Pulo and Bukit Duri to name a few. The two kampungs struggled to assess their right to the city. Unlike the Kampung Pulo case which failed in the class action lawsuit, the DKI Jakarta High Court won the class action lawsuit for residents of Bukit Duri, South Jakarta, regarding the evictions carried out by the Government [16]. This decision strengthens the victory of the residents of Bukit Duri which was decided at the Central Jakarta District Court on October 25, 2017. The victory was however short-lived, but citizens relied heavily on legal aid and advocacy than other commoning practices, showing their determination in saving their own commons.

1.5. NGOs' Role in Commoning

At some cases, NGOs helped informal communities in governing their commons, and Ciliwung Merdeka was one of the NGOs which stands out. The commoning process in Bukit Duri cannot be separated from the Ciliwung Merdeka NGO. This NGO emerged in the year 2000 to transform the kampung environment in overcoming the problem of gangsters, in addition to the participation of the police and religious leaders and local community leaders. With the help of Ciliwung Merdeka, which ever since 2000 has provided studios for learning, dance, music, and theater; residents succeeded in eradicating the negative stigma of Bukit Duri through a process of social organization and communal activities that were more affordable to their environment [16]. Ciliwung Merdeka not only provides studios for community empowerment, especially for village youth, but also provides what is called Padawangi (2018) as the ability to provide village residents with aspirations [16].

Compared to the commoning of Kampung Pulo, which is more aggressive in opposing the eviction plan, the residents of Bukit Duri generally have a more collective and collaborative effort to maintain social relations that have long been sought in the community. Residents assisted by the NGO Ciliwung Merdeka rejected the evictions by also presenting the idea of a Kampung Susun Manusiawi , a 'humane kampung flats' to oppose the old 'exemplary center' and to form a new social housing ideal for the riverbank community. This process illustrates the formal-informal synergies related to Ostrom's ideas on how the commons are managed at a local scale, namely by initiating an 'institutional framework' that defines responsibilities and rights between groups. In this case, representatives of the residents of Bukit Duri through the NGO Ciliwung Merdeka demanded their right to remain in their environment, and they felt they had a responsibility to provide a solution—a flat proposal—as an alternative to solving ecological problem, the annual flooding.

1.6. Commodification of Commons: A Commons' Destruction

At first, commodification and privatization of commons are regarded as a 'correct' way to control and manage the use of commons among members of the community, but this is not the case in recent decades. According to Hardin (1968), privatization of commons can help control the use of resources in it by regulating access to its use. This view is based on the individualistic nature of the commons users, where the commons are considered as something that is not limited so that the commons can be used excessively as a way to accumulate capital to the maximum In other words, privatization as a tool to save the commons is the impact of a paradigm that is still centered on what Lefebvre calls a single power in the form of capital and markets.

Privatization as a response to controlling the use of resources in a commons is opposed by Ostrom (1990) through the view that controlling the use of resources in a commons can be regulated without privatization [2][4]. In some cases, especially small-scale commoning, regulation of resource use in the commons can be done by utilizing existing social capital, such as deliberation and negotiation [2]. The use and development of resources mutually with the existence of commodification and capitalization certainly cannot be considered as means of commoning for the destruction of commons is carried out through commodification and capitalization of commons by the capital or acquisition by the formal sector [1]. Nowadays, privatization is generally not accepted as a solution to govern commons, but rather as a way to destroy it.

However, the accumulation of capital through the commodification of urban commons is not restrained to direct commodification. Urban commons can be habituated even without a direct commodification which makes a commons a commodity. Privatization by the capital and the government or other formal institutions can be a mark of the end of a means commoning that has taken place in a community [1]. The destruction of commons can be inferred from the cases of kampung evictions if the residents could not get a direct benefit from the evictions that were carried out. The privatization of the commons is used as an excuse to save the sustainability of the commons even though in reality capital accumulation continues to occur. This process sacrifices 'the right to the city' especially for marginalized city residents, which in the case of Jakarta are informal citizens. Jakarta's informal communities, especially those living on the riverbanks, often face threats to their city rights in the form of eviction plans in the name of flood control. Kampungs on the banks of the river (which are often referred to as illegal because they were erected on land that extends beyond the river border) experience indirect commodification. What is meant indirectly here is that the evictions are not carried out to commodify the land that is being evicted directly and get benefits in the form of direct capital, the capital will grow around the eviction site as its land value increases, profiting the landlord and not the residents themselves [8]. The increase in land prices around areas that have been evicted for river normalization can indirectly increase the wealth of landowners 'overnight' by selling them, but it may result in gentrification and emptying entire kampung en-masse in the long term (see Padawangi argument from a study in Menteng Atas) [8]. If this happens, the residents will not benefit from the increase in land values. On the contrary, residents would become object of capitalism while at the same time destroying their own commons.

1.7. Commoning as Participatory Solution to Flooding
Flooding is the major issue of kampungs' eviction, and eviction is becoming the 'mainstream' way of dealing with flooding and 'slum' in a single strike. Inspite of this, Padawangi and Douglas highlighted participatory initiatives as other ways of dealing with flooding without the compromise of urban commons that the communities have been building and maintaining for decades. Contrary to the idea that eviction and river normalization is the key to control urban flooding, Padawangi and Douglas (2015) affirms that the existence of an active civil society in the midst of spatial marginalization in relation to environmental degradation and mismanagement opens new political dynamics and is potentially gaining possibilities to address floods along with increasing neighborhood resilience (p. 519) [16]. Using the concept of political ecology, Padawangi and Douglas mapped citizens participatory activities and found that Kampung residents, NGOs, and professionals are "increasingly trying to work collaboratively towards common goals of building flood-disaster-resilient communities" (p. 550) [16]. It does not mean the complete quelling of the current urban flooding, but it does incremental and small-scale initiatives for flooding. To put it in the framework of commons and commoning, residents of informal kampungs might utilize the already mature concept of building and organizing their commons to not only foster their community, but also as a part of participatory solution to urban flooding. This way, both destruction of commons and regular flooding could be handled in just one move, i.e. to utilize their commons for their own and the city's public good.

Commons and Commoning in RW 04 Bidaracina

The eviction involves two RWs (neighbourhood communities) in Bidaracina, specifically RW 04 and RW 14. Bidaracina is a kelurahan located in Jatinegara sub-district of East Jakarta, DKI Jakarta. From among the two RWs, the author interviewed two residents of RW 04 who represented their community in their effort to saving their commons from eviction: Pak Edi as the RW secretary and Mba Astriyani (Mba Aci) as part of the community legal team. To identify commons and commoning in RW 04, the author conducted online interviews with the two residents.

1.8. Class-Action
The main discussion developed from RW 04 Bidaracina focuses on how the recognition of property ownership is obtained through lengthy legal contestations as a form of commoning. Apart from the DKI Jakarta government, another petitioner who sued Bidaracina residents to the cassation level was an individual on behalf of Hengky Saputra. The legal processes to obtain property rights are managed particularly in RW 04, whereas in RW 14 (which is also affected by those cassations) there are no such initiatives, according to Astriyani. The success of the class action initiatives by representatives of Bidaracina residents concluded the resilience of the residents of RW 04. The achievement of the community in fighting for resilience in the form of recognition of ownership of land and structures built

on them can be seen in the successes of residents of RW 04 Bidaracina by the means of class action in obtaining compensation.

The importance of the ongoing legal processes was acknowledged by one of the residents of RW 04 Bidaracina, Pak Edi. He usually calls a resident and advocate named Astriyani, who represents the community at the cassations, as the "chief of the tribe". Such acknowledgment indicates that citizens have more confidence for legal support and advocacy over any other forms of commoning practices. Pak Edi further emphasized the success of the legal team, one of which was Astriyani (Mba Aci) as a resident of RW 04 herself, in educating residents to be more appreciative of their right to the city, hence saved their kampung from a catastrophe like the one happened in Kampung Pulo eviction in 2015, where there was a clash between officials and residents resulted from negotiation failure. This was illustrated in 2015 by how the eviction forces which brought police officers, Satpol PP, fire brigade, and ambulance were able to be restrained because of the cohesiveness of residents who were not provoked by rage and carried out anarchist actions, contrary to what happened in Kampung Pulo which was considered 'too bloody' by Edi.

"Our kampung is safe because there is Mba Aci who has volunteered to become a representative. Thanks to her we can educate the residents here, so they can rest easy and not easily provoked. If the residents are provoked, our kampung will be ruined like Kampung Pulo".

The eviction plan made the residents of RW 04 more integrated: they were more organized, got along well with other residents, and becoming more united, as both Astriyani and Edi felt. Nevertheless, Kampung residents often find the presence of RT (rukun tetangga) and RW (rukun warga) initiatives in their kampung as something they only admit for granted, overlooking their value as a form of resource that provides a livelihood for them. To put it in another way, those initiatives are judged merely as a tool to help them in administration and bureaucracy, not as valuable as what the notion of commons was valued by Hellman (2018) in his research. The invaluableness of commons and commoning in RW 04 Bidaracina, according to Pak Edi, lies saliently in the ease of bureaucracy, prevailing over the importance of building and strengthening social networks among the residents. The Urban commons that they have been building decades might be, in fact, overshadowed by the importance of formal and bureaucracy needs.

"Sometimes we see that the advantages and disadvantages can be non-material material. Non-material values cannot be valued with money. That is hard to judge. ... After all, the effect is that the administration of documents will be difficult, even though the government has said that administrative matters for the evicted will be made easier but we do not know. ... Furthermore, the school children also had to be transferred. Although even when maybe they (the government) will compensate one house to two houses, for example, I still choose to continue to live here"(Edi 2020).

1.9. Incrementality in Kampung's Built Form

RT 10 RW 04 Bidaracina is located on the eastern bank of the Ciliwung river. To enter RT 10 area from Jalan Otto Iskandar Dinata, pedestrians, cyclists, or motorbike can enter via Jalan Sensus Raya, then continue to Jalan Sensus II. The width of the neighborhood alleyway Gang Sensus II D is less than the primary access road, hence not accessible by car or large vehicle. The organization of streets and alleyways are incremental. According to one of the residents, the alleys in RW 04, particularly Gang Sensus II D, are considered as collective properties and have been relatively in the same layout since around the 1960s. Pak Edi considers the "neat, rectangular layout of RW 04" for the notion that Bidaracina has been around for a long time as part of an office that was later abandoned and then reoccupied.

The residents' efforts to secure the tenure of their land and buildings and their efforts to maintain their commons are considered commoning efforts. According to the two respondents, the condition of the kampung in RW 04 Bidaracina as a commons is its overall impression as a 'neat' and 'visually organized' kampung. The condition of the kampung, Edi argued, is not as 'shabby' as other urban

IOP Conf. Series: Earth and Environmental Science **903** (2021) 012005 doi:10.1088/1755-1315/903/1/012005

kampungs that have previously been evicted, while referring to Kampung Melayu, Kampung Pulo, and Bukit Duri. The two residents also felt that their kampung was a neat kampung with expressions such as "it's like a housing complex (suburbia-like development)". Another expression is that their kampung is not composed of totally low-income residents, but rather consists of heterogeneous middle-income residents with varying degree of social status.

"In RW 04, the situation is very heterogeneous. There are residents who live in permanent houses, on roads that can be passed by four-wheeled vehicles, have postgraduate education, and can be classified as middle-incomers. However, there are residents who live in narrow alleys and work in the informal sector"(Aci 2020).

Bidaracina was built incrementally and continuously: a *tropotopia* (see Harjoko 2009 for *tropotopia* in kampung) [18]. Spatial transformations occur continuously in RW 04, which commonly prevails in how the land and buildings adapt to the increasing population. Incremental kampung transformations at RW 04 Bidaracina are carried out by either the hereditary owner or by immigrants. Inheritors played significant role in creating and managing their commons, especially when inheritors are very reluctant to move out of their inherited properties. Instead of moving out of their fathers' land because of the shortage of land, they insisted on dividing it for every inheritor. Families who have lived for three generations or more can identify which land belonged to their family and which was then divided and later built by their grantees (this is extracted form Pak Edi's statement when narrating a story of his neighbor). Due to the inadequate land suitable for the construction of large houses in Bidaracina, some of the grantees have built houses on the already divided inherited land. In doing so, for they are dealing with small patch of land, they had to come with 'creative' ways to build and manage their resources. The reluctance of the inheritors to move outside and the increase in migrants to RW 04 Bidaracina is the impact of the existing commons in the form of a strategic location. Astriyani, as a third-generation middle-high income resident and at the same time a legal advocate of Bidaracina, still lives in his grandfather's house mainly due to its strategic location.

Alongside the increase in population due to the enlargement of families in RW 04, arrivals of other families to RW 04 Bidaracina is also a key factor in shaping the kampung's built environment. Some of the lands inherited to some of the inheritors were then sold to individuals outside of the grantee family. Pak Edi for example, who has been living for 15 years in his residence in RW 04, is one of the immigrants. Pak Edi is one of the families who bought a property in RT 10. The land was bought from a person named Nyi Solihat, where she, interestingly enough, was not the original owner of the property but rather the third hand who lived there. He moved to RW 04 Bidaracina mainly for the proximity to the workplace and relatives, and later made a significant transformation of the kampung's fabric by upgrading his semi-permanent house to a permanent house. He even managed to renovate the house several times 'at the right times'; at times where necessity met financial capability.

The description of Pak Edi's family as part of an 'urban majority', a concept explained by Simone (2014) as the group shaping the image of the city, is understood from its temporal and dynamic nature. Sometimes the 'urban majority' does not do something based on a clear and planned goal or vision [5]. Pak Edi's family as the "urban majority" does something "at the right time". In other words, the development of both commons and private properties are carried out at the right time, with the availability of resources available at corresponding time. Even though this renovation was aimed at meeting the space requirements for his family, Pak Edi did not deny the role of the renovation he did as forming a kampung image that could make the image of his residence in particular and the image of his kampung, to be perceived better. The house he bought from Nyi Solihat, which initially was semi-permanent (with mostly wooden structure and construction), turned into a permanent house (constructed with concrete and bricks). These do not secure a direct effect on the certainty of land tenure, but rather indirectly, resilience is reflected in the 'tidy' environmental conditions: the environment that Padawangi calls an 'exemplary center', an exemplary center in accordance with the spatial vision of the formal sector

IOP Conf. Series: Earth and Environmental Science **903** (2021) 012005 doi:10.1088/1755-1315/903/1/012005

and the societies affected by it. He even admitted that renovation had also increased his prestige and dignity in the presence of not only his neighbors but also from his colleagues at work.

An exemplary center in the form of a neat or "a kampung that resembles housing complex of middle-high class", as being said by Pak Edi, makes people who previously did not realize the complexities of land ownership in the kampung will even becoming more aware of it. The neat appearance of a kampung might "deceive" ordinary people about its status because they will think that a tidy environment is an indicator that a plot of land and building already has a complete certificate of ownership. When they find out that even a tidy and well-organized kampung does not have a title of ownership, residents especially those who dwell in an inherited property from their parents, would question themselves about the status of the land and buildings in which they occupy. This is what Hellman (2018) observed in Kampung Pulo, which the author also observed from Pak Edi's mindset. A settlement without land title records has some points of resilience in its image. The better the built environment of a kampung, including its shared facilities, the better the perceived image of particular kampung will be.

1.10. The Spatial Trade-offs of Kampung's Commons

Musala (communal prayer hall) of RW 04 has been a waqf (benefaction) managed by the residents of RW 04 since the 1960s (Aci 2020, personal interview). Musala is managed by the residents themselves autonomously. Pak Edi, who was also a participant in the interview, was also involved. The area around the musala is also a commons whose use is regulated by the RW or musala committee.

Figure 1. Musala of RW 04

This arrangement for the use of commons is temporal and dynamic. Adjustments to the resource in the form of a prayer room and the surrounding alley occur at certain times. The prayer room is usually filled with worshipers during tarawih in the first week of Ramadan, so alleys that intersect with the prayer room must be temporarily closed. Commoning is observed from the use of musala and part of the kampung alleyway as an overflow extension of musala. Apart from that, musala is also sometimes opened for Friday prayers under certain circumstances. If the musala is opened for Friday prayers, the musala is also usually full and uses two alleys to the north and east of it for use. The use of commons and the switching of commons is manageable and can occur loosely. The residents have tolerated the change in the function of their commons from the use of access (roads/alleys) to social uses, i.e. to worship and congregate. Furthermore, this alternation of commons' function was also recognized and supported by the RW 04 community by providing a canopy covering the alley on two sides of the prayer room that intersected with it. This sort of trade-off occurs because the musala committee and RW 04 residents in general have not been able to divert a large amount of material resources for land acquisition as well as building materials to expand and renovate their musala.

Commons can also be observed from the presence of stall, *warung*, opened by several residents, especially RT 10 RW 04 near Pak Edi's residence. Warung is privately owned, used for profit, and developed from capital that is "planted" in a settlement so that it becomes a shop. However, as the lines between formal-informal dualism in the kampung is not obvious, the position of warung is on the spectrum between the two poles of formal and informal. Although privately owned and not cooperatively, technically a warung has entered the informal sector because it evades and does not pay taxes, among other things (see Portes & Haller 2004, Tunas 2008). Stalls are understood as commons not because of their ownership, but rather spatially. Spatially, the commons are identified as an extension of the house as a private space to the alley in front of it, and vice versa; alley extension in the form of public space to a part of the house. Opening a shop is a manner of competition of urban majority that sees opportunities in their situation. Near Pak Edi's house, there is a neighbor who opens a shop, and the illustrative sentence he said several times in interviews is "he survived by just putting the table in front of the house to support his livelihood". That sentence can be understood as informality that can grow and prosper by means of seeking opportunity out of competitions in a kampung environment.

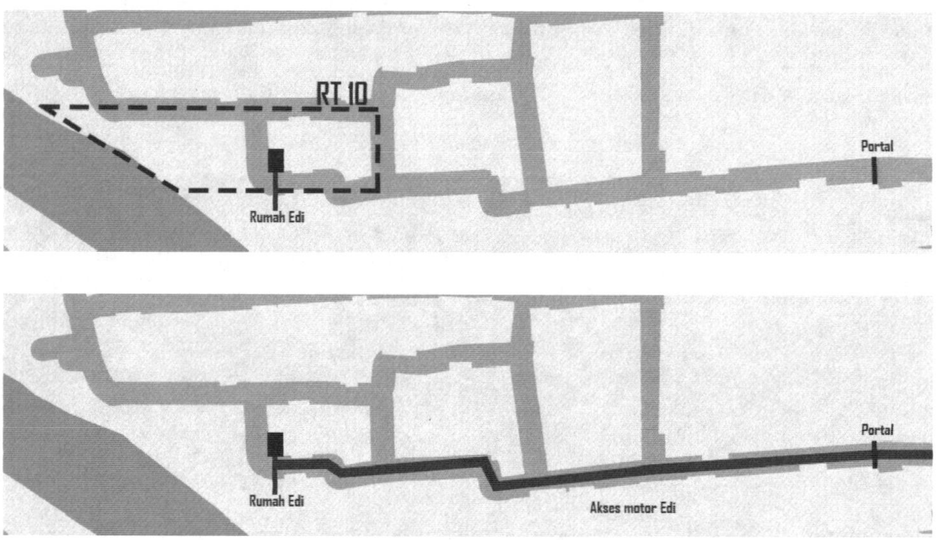

Figure 2 The location of security portal in RW 04

The sense of security can be observed from the kampung level and to a lesser extent, from the level of one family house. Through interviews with Pak Edi, the residents' motorbikes could be parked in front of their respective houses without having to be put inside the house to secure their motorbikes from theft. That is contrary to some cases where residents' motorbikes are usually placed in the house before the house is locked at night, even though the owner of the house has a fenced yard. The habit of storing a motorbike in the living room of the house was not practiced, especially in RW 04. The portal as an additional security element in Bidaracina is only on the main road from the Bidaracina on Jalan Sensus Raya. The portal provides a sense of security for residents of several RWs that have main access from Jalan Sensus Raya to the outside, but not other RT-RT or RW-RW. Teenage brawls from outer RWs have occurred in RW 04 (when a group of teenagers entered through the portal before the portal was closed), but brawls between RTs inside RW 04 and surrounding RWs that have access from Jalan Sensus Raya, according to Pak Edi, has never happened. In other words, streets and alleys inside RW 04 were not ported because the residents perceived that their environment was considered safe and that their relationships were excellent so that residents did not mistrust each other.

1.11. The Destruction of Bidaracina's Commons

To understand that eviction is a form of capital accumulation, we have to look into why Bidaracina eviction is planned in the first place. The underlying motive for Bidaracina eviction as an effort to control floods is, arguably, one of the steps to provide an indirect capital stimulus as part of a continuous effort so that the capitalist system continues. The eviction of Bidaracina was intended to improve the flow of Ciliwung river when the volume of flow was unusually larger by channeling it to BKT [9], which would not only minimize losses due to flooding in downstream areas but would also increase land prices around the riverbank by minimizing the possibility of great urban flood. This is an indirect control of the capital surplus, using the framework by which Harvey developed. The alliance of the state with capital and the 'creative destruction' of the market creates new space for capitalist expansion and economic growth. Bidaracina evictions are not aimed at directly creating capitalist developments for development projects, but this kind of eviction practice "works by changing the identity attached to a place and thereby increasing the value of land in the surrounding environment (see Leitner and Sheppard 2018) ". In other words, this eviction will 'open the gate' for further capitalist development around the affected area, hence will further disrupt the continuous process of commoning that residents built over the decades. Commons will be lost.

Although eviction (which means capital accumulation and further destruction of commons) in dealing with urban flooding seems impossible to go by the board, the last resort would be to utilize RW 04 own commons to save themselves from the destruction of their commons as what was discussed in section 3. Participatory initiatives, as what Padawangi and Douglas argued, residents of informal kampungs might utilize the already mature concept of building and organizing their commons to not only foster their community, but also as a part of participatory solution to urban flooding [16]. To quell the eviction to save RW 04 as a whole commons altogether might be impossible, however, it is very possible that a different kind of commons might still be intact: the social and cultural resources in the form of community initiations between residents. Kampung residents and soon-to-be former residents, NGOs, and professionals can work collaboratively towards common goals of building resilient communities.

As the certainty of when would eviction and compensation take place remains unclear, as it is perceived by Edi and Astriyani, notwithstanding the increasingly inevitable evictions marked by an appraisal process for compensation that has been planned by the beginning of 2020, commoning will likely end by eviction and displacement. Although there is a possibility where some residents will stay in their old house because they are only partially affected, Bidaracina residents who have been completely evicted will only enjoy very brief and temporary resilience. After the eviction was carried out, the resilience shown by the residents was in the form of recognition of land ownership would be ended in capital accumulation in the form of evictions for the flood prevention project. The land where they live, which Pak Edi described as "invaluable by money" has replaced its meaning into something else, namely something that provides a stimulus for the continuity of the capitalist order under the term of flood control.

Conclusion

Despite the limited spatial data that can be collected from the research at RW 04 Bidaracina, the findings are consistent with what has existed in existing studies. Several things can be identified as commons: resources that can be enjoyed physically and non-physically. From the case of RW 04 Bidaracina, the means of commoning is observed from the incrementality both in the formation of their residential circumstances, initiating a business, the formation of spaces for worship, and the form of streets or alleys. Commoning in Bidaracina is essentially all the strategies undertaken by the residents for the development and use of both physical and non-physical resources as livelihood strategy, but that these practices must contribute to the communal benefits of the kampung. The limitation to commoning lies in the privatization/commodification of common resources, of which eviction is one of them. Commoning that existed in kampungs before an eviction can be recognized in different events and highly dependent on other institutions involved in them. The end of their commoning culminates in the

IOP Conf. Series: Earth and Environmental Science **903** (2021) 012005 doi:10.1088/1755-1315/903/1/012005

destruction of the commons which are generally due to the privatization and commodification of the commons, especially through evictions.

References

[1] Gidwani V and Baviskar A 2011 Urban commons *Econ. Polit. Wkly.* **46**(50) pp 42-3
[2] Harvey D 2012 *Rebel cities: from the right to the city to urban revolution* (London: Verso)
[3] Hardin G 1968 The tragedy of the commons *Science* **162**(3859) pp 1243-48
[4] Ostrom E 1990 *Governing the Commons* (Cambridge: Cambridge University Press)
[5] Simone A 2014 *Jakarta: Drawing the City Near* (Minneapolis: University of Minnesota Press)
[6] Hellman J 2018 How to prove you are not a squatter appropriating space and marking presence in Jakarta, ed Cabannes Y, Douglass M, and Padawangi R 2018 *Cities by and for the People in Asia* (Amsterdam: University of Amsterdam Press)
[7] Dovey K, Cook B and Achmadi A 2019 Contested riverscapes in Jakarta: Flooding, forced eviction and urban image *Space Polity* **23**(3) pp 265-82
[8] Padawangi R 2019 Forced evictions, spatial (un)certainties and the making of exemplary centres in Indonesia *Asia Pac. Viewp.* **60**(1) pp. 65-79
[9] Yunita R et al. 2017 Kajian aliran inlet Sudetan Sungai Ciliwung ke Kanal Banjir Timur untuk pengendalian banjir Jakarta. Malang: Fakultas Teknik Universitas Brawijaya
[10] Snyder H 2019 Literature review as a research methodology: An overview and guidelines *J. Bus. Res.* **104** pp 333–9.
[11] Leitner, H. and Sheppard, E. (2017). From kampungs to condos? Contested accumulations through displacement in Jakarta *Environment and Planning* A **50**(2) pp 437-456
[12] Jones P 2017 Housing resilience and the informal city *J. Reg. City Plan.* **28**(2) pp 129-39
[13] Betteridge B and Webb S 2019 Everyday resilience, reworking, and resistance in North Jakarta's kampungs *Nat. Space* **0**(0) pp 1-23
[14] Maring P, Hasugian F and Kaligis R 2015 The portrait of social strategy of Ciliwung riverbank community. *J. Komunitas* **7**(1) pp 102-11
[15] Sari D, Ramadhonah R, Innaqa S, Sugiana A, Irawan D, Ratnadewi, Supriyono and Harmanto D 2018 Obstacle and driving factors of Ciliwung river revitalization and community acceptance of the new land use *Int. J. Eng. Technol.* **7**(2.14) pp 256-9
[16] Padawangi R 2018 Excavating the ruins of aspirational urban futures in Bukit Duri, Jakarta. In Bunnell T and D.P.S. Goh (eds.), *Urban Asias: Essays of futurity Past and Present* (Berlin: Jovis)
[17] Padawangi R and Douglass M 2015 Water, Water Everywhere: Toward Participatory Solutions to Chronic Urban Flooding in Jakarta *Pac. Affairs* **88**(3) pp 517-50
[18] Harjoko T 2009 *Urban Kampung: Its Genesis and Transformation into Metropolis, with Particular Reference to Penggilingan in Jakarta* (Saarbrucken: VDM Verlag Dr. Müller)
[19] Tunas D 2008 *The Spatial Economy in the Urban Informal Settlements* (Amsterdam: International Forum on Urbanism)

Acknowledgments

We acknowledge generous funding from Hibah PUTI (Publikasi Terindeks Internasional) No: NKB-1063/UN2.RST/HKP.05.00/2020. We thank Mba Astriyani "Aci" Ahmad and Pak Edi Hermadi for the consideration of their time spent for the interviews.

Environmental Concern to save Langgar Tinggi Heritage Mosque at Kampung Pekojan

Gianoora Achmad[1] **and Kemas Ridwan Kurniawan**[1]

[1] Architecture Department, Faculty of Engineering, University of Indonesia, Depok 16424, Indonesia

Email: gianoora.achmad@ui.ac.id

Abstract. Located in Tambora, West Jakarta, Kampung Pekojan has an essential role as a part of historical stories back to the colonialism era in Batavia, now known as Jakarta. As the residential area of Muslim for the Arabs, The Indian, and the indigenous people, Pekojan has six cultural heritage mosques, one of them is Langgar Tinggi. However, the yearly flood event eventually become a problem and it endangered Langgar Tinggi and its surrounding. Using descriptive qualitative methods by collecting data which includes literature studies, interviews with the experts, this paper is expected to provide the necessary information and data about how to handle a flood event around Kampung Pekojan and save Langgar Tinggi mosque as an important cultural heritage and a place that holds history.

Keywords: Pekojan, Langgar Tinggi, Mosque

1. Introduction

Located in Tambora, West Jakarta, Kampung Pekojan is a sub-district which has an interesting history background of how the *Kampung* became an important part that formed Batavia– now Jakarta. Kampung Pekojan is a sub-district located in Tambora, West Jakarta which has an important role in historical story of colonialism era when Jakarta was still called Batavia. During the colonial period, Pekojan was a special residential area for Muslim and Indian descent communities who were Muslim. The Arabs who came to Indonesia at that time had the aim of trading and preaching such as teaching the Koran, building boarding schools, and mosques.

Six mosques become Pekojan's cultural heritage, and they should be protected as the mosques themselves are a significant part of Kampung Pekojan's community history and the beginning of the spread of Islam in Batavia at that time. But as we know, Jakarta is a complex urban area with socio-economic problems and indirectly triggering a flood event. The flood is the yearly problem for Pekojan even since colonialism. Local government responds this by increasing the elevation of the road every year, on the other hand, this way out to fix the problem but also it becomes an issue since it makes one of the mosques, Langgar Tinggi and some of the settlement area vulnerable because their location is lower than the road.

IOP Conf. Series: Earth and Environmental Science **903** (2021) 012006 doi:10.1088/1755-1315/903/1/012006

2. Methods

For this research, descriptive qualitative used as the primary method. Primer data gathered by collecting literature studies to find suitable theories to support the arguments. Analysis in this research is based on the observation of the site and interviews with experts and native people of Kampung Pekojan. Secondary data gathered by collecting data from the official Indonesian Government site and other accountable sites.

To be noted that this research is done under the condition of the Covid-19 situation. Interviews were done by messages exchange and phone calls, and an observation of the site was done months before pandemic Covid-19 happens.

3. Historical Background

3.1 Flood History in Jakarta

Jakarta area is a complex urban area with socio-economic problems and indirectly triggering a flood event since colonialism era. On Tugu inscription from Tarumanegara Kingdom, declared about a flood and its response in the fifth century AD.

Floods that occur in an area are closely related to natural conditions and urban planning. Natural conditions related to geographic factors such as geology, climate, rainfall, and geomorphology are elements that change slowly or even hundreds of years. In addition to geographical factors, another factor influencing the size of flooding is Jakarta's urban planning policy which includes city development and the behaviour of its inhabitants.

Development carried out in Jakarta has rapidly transformed natural ecosystems into artificial ecosystems. It increased rapidly from 1970. The area of wet land and water bodies in North Jakarta in 1970 was 2,620.22 hectares. In 1980, the wet land area was reduced to 2,343.55 hectares and in 1990 the land was reduced to 698.23 hectares. Reduction of actual wet land area unavailable for building use because it functions as a water house during the rainy season results in a reduction in the area and number of water retention locations during floods while a reduction around agricultural land use and open space results in a decrease of the ability to absorb or infiltrate water into soil.

Not only was 1918 a matter of debate at the city council at the time, but it is also said that the flood of 1918 was the biggest in the last 20 years. In the rainy season of 1817, the rain fell even longer, said to be 40 days of continuous rain so that the roof tiles were neglected and broken, but the floods that came in Batavia were not as big as the 1918 floods. In the 1918 floods, the flooding area that hit Jakarta experienced an expansion to westward, the area around the Kali Angke, Pekojan Pejagalan, began to be flooded. For10 years the flood has become more widespread. Kampung Pejagalan, West Jakarta, also called as a Kampung Air. This naming is related to the condition of the village which when the rainy season is always submerged in water. The Angke River is an important part of the process of storing and flowing water.

After the 1918 floods, Jakarta was hit again by floods in 1919. On 5 and 6 February 1919, high rainfall caused flooding in *Weltevreden* (the main residential area of the Europeans on the outskirts of Batavia, the Dutch East Indies which is approximately 10 kilometres from old Batavia to the south). There was also flood in 1923 but was not too massive.

Jakarta Special District area itself is 662 km^2 of land area and 6,977 km^2 of sea area. On the northern part of Jakarta, the coast stretches along 35 km, 13 rivers and 2 canals are disemboguing into the coast.

With that condition, as explained above, even until now Kampung Pekojan, West Jakarta, where Langgar Tinggi stands, is one that has the potential to get annual floods every year.

3.2 History of Langgar Tinggi Mosque

The mosque is located between Pekojan street across Angke River, Tambora subdistrict, West Jakarta. The coordinate is 6°08'29.54"S 106°48'21.38"E/ -6.14154, 106.80594, no changes of location from the original since the first time the mosque founded. The Langgar Tinggi was built around 1830 by Muslim trader from Yaman named Abubakar Shihab on donated land of *waqf* (religious endowment) of Sharifah Masad Babrik Baalwi. She was a family member of a wealthy Palembang family, and they left her massive inheritance to be administered as a waqf by Shaykh Said bin Salim Naum. The latter, who was appointed headman of Arabs in Batavia in 1849, donated land from the waqf for the purposes of building the Langgar Tinggi Mosque and a Muslim cemetery in the Tanah Abang district. At that time, wealthy traders donated land or buildings for charitable purposes in the form of waqf for the establishment of facilities for Muslims.[1] Built as a two-story building, the essential values of the architecture of Langgar Tinggi Mosque can be seen from its mixture elements from several cultures (eclectic architecture style) such as Chinese, European, and Javanese architecture. This architectural style is considered quite typical of the architecture of Java.

Figure 1. Langgar Tinggi Mosque and Angke River circa 1900
source: Royal Netherlands Institute of Southeast Asian and Caribbean Studies and Leiden University Library/Wikimedia Commons

In the nineteenth century, and the variations of the same may be found in Cirebon, Pasuruan, and Gresik. Within the Langgar Tinggi mosque, a lacquer work from, and Palembang decorated the *mihrab* (the niche indicating the direction of Mecca). Langgar Tinggi is also a legacy of an eventful shift from Palembang to Batavia (Sumit K. Mandal, 2018: 72).

The Langgar Tinggi previously was integrated with Angke River, so that many traders and boatmen could directly take *wudu* (the Islamics procedure for cleansing parts of the body, a type of ritual purification, or ablution) from the water of Angke River. At that time, every Thursday night there was the reading of *mawlid* (birthday of the Islamic Prophet Muhammad) and every Monday night there was the reading of *Al-Burda* (a thirteenth-century ode of praise for the Islamic prophet Muhammad composed by the eminent Sufi mystic, Iman al-Busiri of Egypt) accompanied by *Rebana* (Malay tambourine that is used in Islamic devotional music in Southeast Asia). The building often used to gather and perform prayers simply; the upper floor used as a place to pray, and the lower floor was occupied by families to live and trade. The Langgar Tinggi was not used for Friday prayers, so the Arabs held Friday prayers at a large mosque, located near the Langgar Tinggi called An-Nawier mosque.[2]

4. Analysis and Finding

4.1 Current Situation of Langgar Tinggi

Almost 190 years after, in the year of 2020, The Langgar Tinggi mosque still standing, and the building managed by Pengurus Langgar Tinggi Pekojan. The building area is 251 m^2, consisting of two floors; the upper floor now still used as a mosque for daily prayer while the lower floor is for the residence of mosque caretaker and space for shops.

Figure 2. Langgar Tinggi Mosque in 1949 source: Collectie Stichting Nationaal Museum van Wereldculturen

Figure 3. Langgar Tinggi Mosque in 2020 source: personal documentation taken January 16th, 2020

In 2019, *Balai Pelestarian Cagar Budaya Banten* (Banten Cultural Heritage Preservation Office) did the revitalization of The Langgar Tinggi Mosque. There is a sign on the front door of the building about the first year of the construction said in *Hijri* (The Islamic Calendar, a lunar calendar consisting of 12 lunar months in a year of 3554 or 355 days) and AC. But the equivalent is incorrect; 1249 H is 1833 AC, not 1829 AC. As explained above, The Langgar Tinggi was built sometime around 1830 (Sumit K.Mandal, 2018: 72), as an essential part of the history of Kampung Pekojan and many people do come to the building like students, scholar, and researcher to learn, it is better to fix and replace the sign to avoid confusion in understanding its history.

Figure 4. Sign on the front door of Langgar Tinggi source: personal documentation, taken February 11st 2020

People tend to call the building with "mosque" after the original name, which is "langgar", while the original meaning is slightly different, and there is an overlap from each meaning. Based on The Great Dictionary of the Indonesian Language of the Language Centre online, an official dictionary produced by the Language Centre of the Indonesian Ministry of Education and Culture, *"langgar"* means *"masjid kecil tempat mengaji atau bersalat, tetapi tidak digunakan untuk salat Jumat"* (a small mosque to read the Quran or praying, but it is not used for Friday prayers). In the meantime, people call the building with *"masjid"* (mosque). In contrast, the meaning of *"masjid"* is *"rumah atau bangunan tempat beribadah orang Islam: setiap Jumat dilakukan salat bersama" (home or building where Muslim's worship: there is joint prayer every Friday).*

IOP Conf. Series: Earth and Environmental Science **903** (2021) 012006 doi:10.1088/1755-1315/903/1/012006

The building as a *"langgar"*, by Ministry of Education and Culture of the Republic of Indonesia, the Langgar Tinggi already recognized as cultural heritage by registered number RNCB.19930329.01.000758 and decision letter, *SK Gubernur No. 475 Tahun 1993.* But the building as a *"masjid"* has not recognized as cultural heritage; based on the official site of Ministry of Education and Culture of the Republic of Indonesia, the registration date of the building was February 1st, 2019, and the status still on verification process by the local government.

4.2 Langgar Tinggi and Environmental Change

As explained above that the Langgar Tinggi is in the potential area of a flood event and from observation to the site, the local government responds this problem by increasing the elevation of the road every year, on the other hand, this way out becomes an issue since it makes Langgar Tinggi building and some of the settlement area around vulnerable because now their location is lower than the road. This statement also came from the interview with the manager of Langgar Tinggi, Bapak Habib Ahmad Alwi Assegaf, "the government did this every year to fix the road (increasing the elevation of the road)." He spoke. These issues become a never-ending circle: flood event comes every year then the road around Langgar Tinggi Mosque exposed by the water for a long time. It makes them damaged and dangerous for people to past. The government repairs the road by pile up the damaged road with new asphalt and it makes the road higher than before.

Picture 5. Langgar Tinggi, year unknown source: Official site of DKI Jakarta province

Picture 6. Langgar Tinggi, year 2016 source: Official site of DKI Jakarta province

IOP Conf. Series: Earth and Environmental Science **903** (2021) 012006 doi:10.1088/1755-1315/903/1/012006

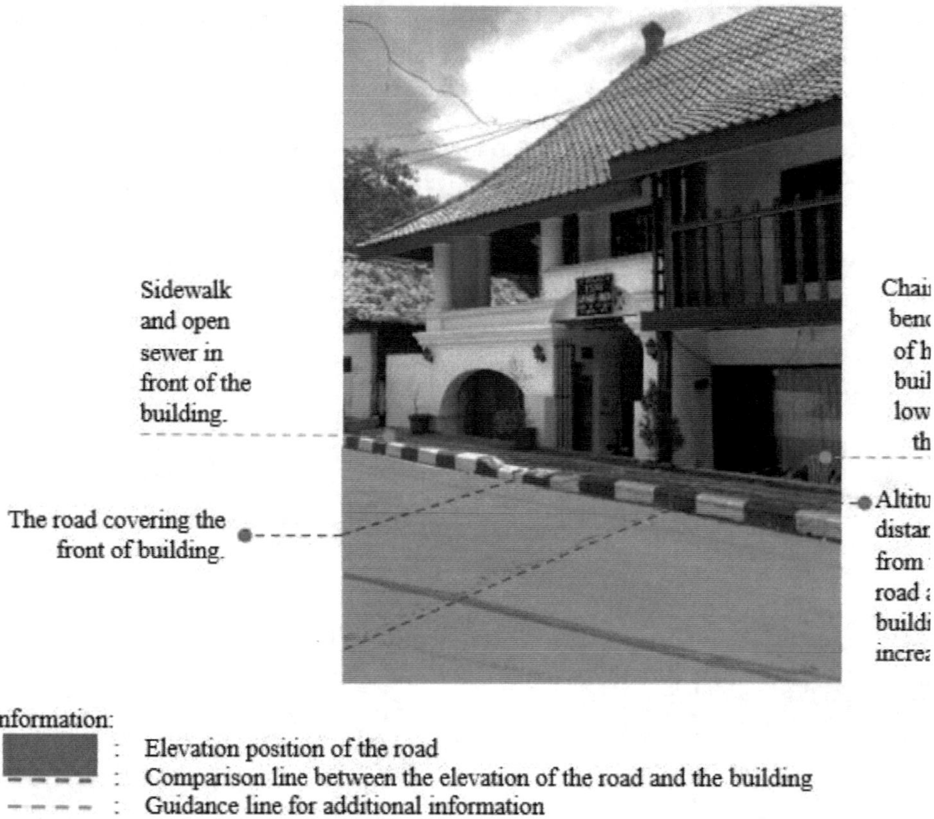

Picture 7. Langgar Tinggi, year 2020 source: personal documentation, taken February 11st 2020

5. Recommendation

First, the related government needs to re-review the methods and how to repair the road every time or every year without increasing the elevation of the road.

Second, optimized the function of the open sewer in front of Langgar Tinggi Mosque (figure 7)

Third, community participation with each other and related government is needed.

Fourth, the verification status of Langgar Tinggi Mosque on the official site of the cultural heritage list produced by the Ministry of Education and Culture of the Republic of Indonesia is still on the process even the building already recognized as cultural heritage in 1993. They need to make an official statement regarding the issue to avoid public confusion.

6. Conclusion

Langgar Tinggi, as one of the oldest mosques in Jakarta, is an essential part of the history of how Jakarta formed from the colonialism era until now; even the community around the building tells the tale if anyone comes intending to understand the story behind 190 years old building. With all its essential values, it is necessary to put the building under protection that comes from an awareness of the community and the government. Currently, the related government already made several moves such as revitalization in 2019, Langgar Tinggi already recognized as cultural heritage in 1993– even they need to make it clear about the building status (see the recommendation, point number four), and how they repair the road in front of the building every year. It needs to be understood very well that Jakarta also has a history of flood events.

Flood has always been a problem every year in Jakarta, and Langgar Tinggi Mosque is in the potential area of flood events. With that information, Kampung Pekojan's community and the related government should more aware and ready to face the event. As explained above, the higher the elevation of the road,

IOP Conf. Series: Earth and Environmental Science **903** (2021) 012006 doi:10.1088/1755-1315/903/1/012006

it is a way out to repair the road that damaged by the flood, but at the same time, it becomes a problem when the face of Langgar Tinggi Mosque covered. In the future, it does not need to change everything because what had been done is done. Re-review optimized installed feature around Langgar Tinggi Mosque. So, the building is not just a protected place that stands silently, Langgar Tinggi Mosque needs to be seen. Because that is how we respect appreciate a historical place.

Notes

[1] Based on *Becoming Arab: Creole Histories and Modern Identity in the Malay World* by Sumit K. Mandal, page 72 and see the official site of Ministry of Education and Culture of the Republic of Indonesia

[2] Based on an interview with Bapak Usman from Kampung Pekojan

Bibliography

[1] Mandal Sumit K 2018 *Becoming Arab: Creole Histories and Modern Identity in the Malay World* (United Kingdom: Cambridge University Press) p 72
[2] Nicholson David 2009. *Environmental Dispute Resolution in Indonesia.* Brill, p 189
[3] Gunawan, Restu. 2010 *Gagalnya Sistem Kanal: Pengendalian Banjir Jakarta dari masa ke masa* (Jakarta: PT. Kompas Media Nusantara) p 4,19,26,128,129
[4] Usman 2019 *Menggali Kembali Kisah Sejarah, Budaya, Para Tokoh, dan Bangunan Cagar Budaya di Kampung Arab Pekojan.*
[5] www.cagarbudaya.kemendikbud.go.id accessed on June 18th, 2020
[6] www.jakarta.go.id accessed on June 18th, 2020

Interviewees:

Bapak Usman as native people of Kampung Pekojan

Bapak Habib Ahmad Alwi Assegaf as the descendant of the Arabs and a manager of Langgar Tinggi Pekojan

Identification of the Architectural Housing Form and the Existence of Ulos Communities in Hutaraja Village, Samosir, North Sumatra, Indonesia

R R Tobing[1] and A K Sakti[1]

[1]Department of Architecture of Universitas Katolik Parahyangan

rumi@unpar.ac.id

Abstract. Tourism has played a major role in boosting the economy of every country in the world. Nonetheless, the COVID-19 pandemic has obviously taken its toll since it hit Indonesia over a year ago. This condition has urged the Indonesian government to diligently make optimal changes, especially in improving regions that have been neglected this whole time, with the purpose of establishing a tourist destination that will prove interesting to tourists. Parahyangan Catholic University plays a role in providing information in the form of scientific studies regarding architecture of the traditional Batak Toba habitation in the Samosir regency of North Sumatera province. The Hutaraja village has been selected as the object of research with the foundation being the existence of *ulos* fabric artisan communities as an appealing feature. This *ulos* fabric plays an important role in the cultural lives of the Toba Bataknese, and may have potential in value and aesthetic uniqueness. The purpose of this study is to identify the relation between the architectural housing form and the existence of the ulos fabric artisan communities. The study has been conducted through a qualitative approach with exploratory and explanatory methods. Its findings show that the Hutaraja village's housing form is strongly related to both cultural and environmental aspects. On the other hand, the existence of *ulos* fabric artisan communities has been discovered to change the attributes of interior and exterior housing forms.

Keywords: housing form, ulos weaver, Hutaraja village

1. Introduction

The tourism sector has played an essential role in boosting the country's economy, even though the Covid-19 pandemic has obviously taken its toll for more than a year. However, various development programs have still been run in different areas to increase the tourism sector. Following the national program, Parahyangan Catholic University has also provided input through scientific studies in architecture that investigated Batak traditional housing in the Samosir District, focusing on its tourism. The research has been conducted based on various aspects, including the lay-out, accessibility, activity pattern of the inhabitants, and other aspects (both physical and non-physical) that influenced the housing space in supporting the tourism activities. This scientific study is expected to give input, suggestions, consideration and evaluation for the local government in developing the Lake Toba area into a tourist destination.

Huta Raja village is one of the settlements in the Lake Toba area selected for a reorganization project to become one of the best destinations to support the government program in developing tourism. Huta

Raja village is also known as a traditional Toba Bataknese housing form that still exists in modern life, with the Community of *Ulos* Traditional Clothing Weavers. This community became the main reason for the existence of Huta Raja village that eventually captured the central government's interest in developing tourism in the area [1].

The research on the physical lay-out of the Toba Bataknese traditional housing architecture based on the existence of the Community of *Ulos* Traditional Clothing Weavers is fundamental because by investigating the influence of non-physical factors of the community on the lay-out of the village, we can help the revitalization process of the traditional village to be transformed into a tourist destination. This study aims to identify the relation between the housing lay-out and community existence in Huta Raja village. This research is expected to contribute both in theory and practice for the community, especially for the village council and local government, toward tourism development in the area. The results of this research are expected to benefit the revitalization process of Huta Raja village in Lumban Suhi-Suhi Toruan with the suitable empowerment of the community. This condition is a positive influence for the development of the physical settlement environment, especially in the face of climate change.

2. Methods

This study uses a qualitative approach with the analytical descriptive method (Bungin, 2004). Data analysis has been made in the data description phase to find the tendency based on this set of data to analyse its significance and relevance. After the complete data had been collected, the set was identified based on the primary factors that form housing (such as culture) that would subsequently be analysed in greater depth through the elements that formed the culture in housing. These elements were identified, followed by the analysis process to establish the roles of each element in the culture as the primary factors that formed the housing lay-out, in this case of the Toba Bataknese traditional housing compounds.

This was followed by analyzing the use of space for *ulos* weaving activities, as a result of identification. The research time was allocated as follows. Phase 1: Researchers conducted a field visit in February 2020. Phase 2: The research was continued by a local team in June-December 2020, due to pandemic conditions, so they were unable visit the study object. The research location is in Lumban Suhi-Suhi Toruan Village, Pangururan District, Samosir Regency, North Sumatra, and the exact location is Kampung 'Ulos' Huta Raja, as can be seen in figure 1.

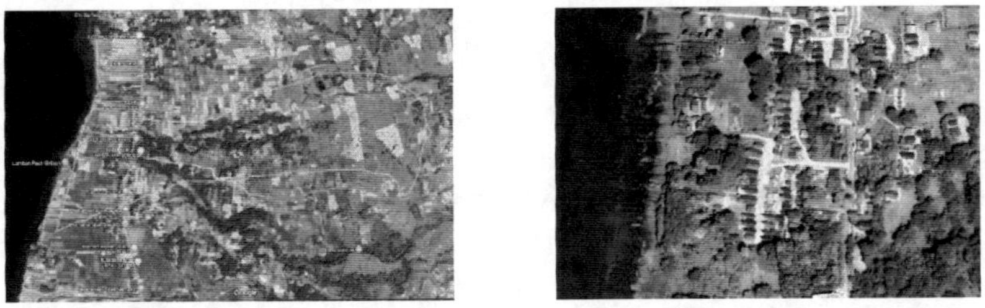

Figure 1. Location of Kampung Huta Raja

3. Results and Discussions

Three basic references have been used in this research.

3.1. Architecture of housing lay-out

Amos Rapoport stated that architecture is the space for humans to live that is seen physically and is strongly related to the essential culture and norms of a group of people. These norms include social and cultural norms that will influence the architectural form. Two factors can influence the housing forms: (1) Primary factors (socio-cultural factors) and (2) secondary factors (modifying factors). Primary factors that influenced the physical form process include culture, while other factors such as climate, geographical location and condition, as well as politics and economics are included in modifying factors See Figure 2.

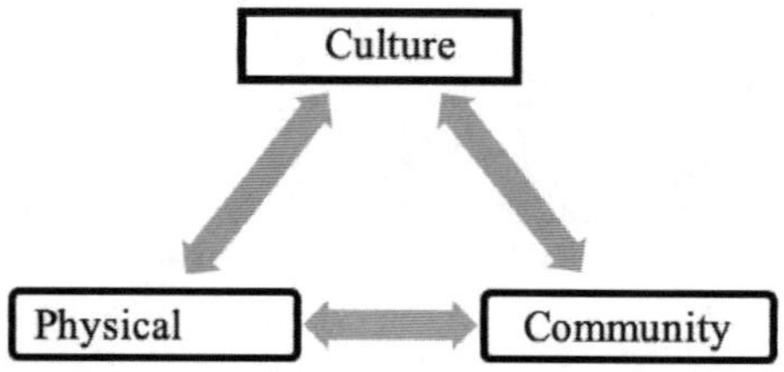

Figure 2. Relations in the Housing Form Process

Rapoport has identified several elements that will form a culture inside a housing, namely: (1) Location, which is the physical geographical location of a housing; (2) Relations to the surrounding nature, where there is landscape with certain functions based on vegetation, natural surroundings, and topography; (3) The presence of physical elements that become the characteristics of the housing; (4) A specific housing pattern based on the culture of the community for specific purposes, such as familial relationship, being nature-oriented, religion, and others; (5) Rooms or spaces with specific types based on their functions or usage, like traditional ceremonies, funerals, public spaces, and others; (6) Specific names for all physical elements based on the local culture; (7) Using a specific orientation system, as the reference for physical housing development according to the living pattern of the society; (8) Colors, textures, and specific ornaments as a part of physical characters; (9) Sounds, smells, temperature, and specific movements, which became unseen characteristics that can increase the sense awareness of the inhabitants; and (10) Different groups in society with specific activities that will attract the attention of the public outside of the housing with their activities [2].

3.2. Toba Bataknese traditional housing architecture

The lay-out of Toba housing usually formed a linear pattern, which is considered whole and strong according to the ancestors. There is a square in the middle of the two linear buildings, in a *huta* (village) that is called *alaman* [3]. The area of a *huta* is usually shaped like a square and bordered by a two-meter tall stone fence with sharp bamboo on the top, called a *parik,* as the protector from human enemies and wild animals. At a glance, the shape of the village/*huta* is like a fortress [4]. See Figure 3.

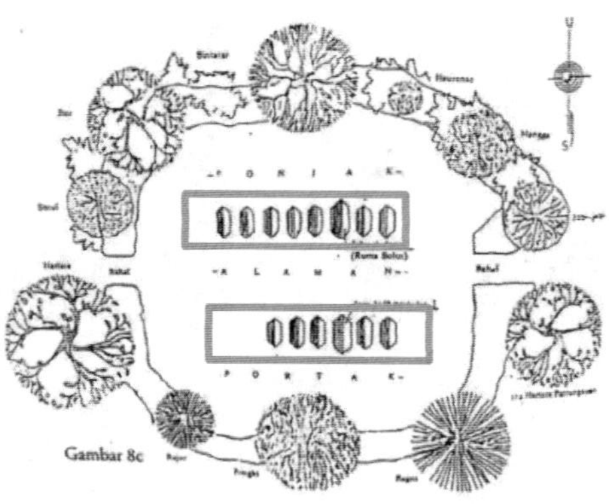

Figure 3. The Concept of *Huta* Layout

(Source: Arsitektur Tradisional Sumatera Utara, 1997)

The Toba Bataknese traditional building has an important role as a set of living guidelines to the community, the symbol of microcosmos that is divided into the three parts or *Tri Tunggal Banua*, which consist of *Banua Toru* (under the Earth) for the leg of the house, *Banua Tonga* (Earth) for the body of the house, and *Banua Ginjang* (the lion in the sky) for the roof of the house [5]. To Bataknese society, the respect for the microcosmos (power of nature) is placed at the top in the hierarchical ranks of their beliefs, long before Christianity entered the Batak land. The traditional Bataks believed in the concept of creation by *Mulajadi Na Bolon* who lived beyond the sky, and had a name according to his position [6].

The shapes of Toba Bataknese traditional buildings are quite similar to one another, which made it difficult to differentiate between them in the existence of housing space. Domenig (2003) conducted a comparison study between several building constructions and concluded that the characteristics of the housing's structural system or *jabu* was a "heritage" or a copy of *sopo*. This happened because Batak Toba tribe had an old traditional of changing the *sopo* that used to be a rice silo into the housing or *jabu* if the number of inhabitants was increasing [7].

3.3.Architecture in the community context

According to Sasongko (2005), a community can be defined as a group of people (togetherness) that share things in their lives [8]. According to Crow and Allan (1994), a community can be divided into three components: based on their location/place; based on their interests; and based on the certain ideas that they share [9]. Meanwhile, Habraken in Wikantiyoso (1997:27) stated that as a product of community, the shape of housing space is created based on the social agreement, and not on an individual product [10]. Therefore, different communities will have different housing characteristics. The design of a house and housing shows the relationships between its individuals, family, and community, which will depend on each local culture. The consequence of this is that the spatial organization in a house, the lay-out of housing, and the accessibility to public facilities will be influenced by the community's outlook on life. Based on the ratio of space that will be used for living and for doing crafting activities for the economy, (Silas, Johan, 2000: 233), the pattern of a housing space is divided into three types [11]:

- Mixed type, where a house is also used as a work space, but the main role of the house is as a living space. The two spaces have the same accessibility.
- Balanced type, where a living space is separated by the communal activity/working space in the same mass. In this type, there is similar interest between the housing as a living space and working space. Accessibility to the housing space is different from the working space.
- Separated type, where a working space is more dominant than the living space, so that the living space is located separately with the working space mass. The accessibility to the living space is different from the working space, as can be seen in Figure 4.

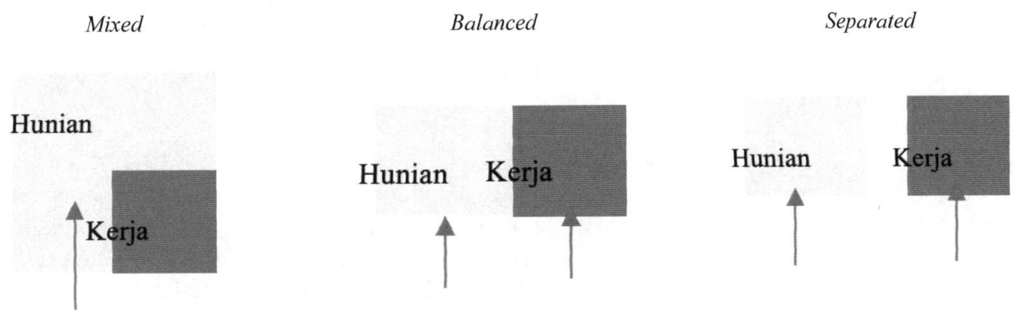

Figure 4. The Concept of *Huta* Layout

N.B. *Hunian* (Yellow) = Living Space & *Kerja* (Red) = Working Space

(Source: Arsitektur Tradisional Sumatera Utara, 1997)

3.4. Identification in the field of architecture

The term and concept of identification in the realm of architectural theory stand for the attachments between the tangible (*kertaji*) aspects, that is to say real, concrete things that can be seen and felt in works of architecture. On the other hand, intangible (*akertaji*) aspects refer to abstract matters such as function, culture, behavior and the like. This can be seen in: structural elements (elements); experimental (space organizations); architectural elements (elements in space which accommodate the activities). The social aspects of function that are represented in the form of objects and space will follow. When one thinks about an object or specific space, one usually thinks of the meaning related to it; that is, what one can do or associate with such an object or space. [12]. The close relationship between function and cultural aspect will be shown in the architectural spaces dealing with both physical and traditional aspects of activities in daily lives. Identification in architecture can be classified into three groups, namely: form, function and culture.

4. Findings

The primary factor in forming the "physical form" of Huta Raja village is the culture, while other factors like climate, geographical position and condition, politics and economics are factors whose influence cannot be ignored. Based on the primary factor, the main discussion of the study objective will identify the elements that formed the housing related to the culture in Huta Raja village, in line with the housing theory as the implementation of the culture (according to Rapoport) that will have findings on the existence of the elements and their roles in forming the lay-out of traditional housing in Huta Raja village. The second objective is to identify the activity space of the weaver community that created the traditional cloth of *ulos* in Huta Raja traditional housing, in relation to their existence as a cultural group that formed the characteristics of housing space. The identification of activity space resulted in findings of the patterns of housing space in the lay-out of Huta Raja traditional housing.

4.1. Elements that formed the culture in Huta Raja Village

The elements that formed the culture in Huta Raja village can be identified geographically as the landscape concept, which consists of:

- Space lay-out is determined by the physical form of the area that has physical borders.
- The condition around Huta Raja village based on its landscape is as follows: landscape determined the function of building mass and circulation pattern according to the needs of the community; vegetation is located within the scope of housing space; mass dominance is located in the relatively flat area; there is no elevation difference in the housing area, so it will make the circulation flow easier.
- Physical elements in Huta Raja village consist of: accessibility that is formed based on the axial path where the recipient space is not influenced by access position; main circulation in the village area is for private use or purposes and has linear pattern where the building mass is facing the main circulation directly.

4.2. Pattern of physical housing space

According to the tradition of Toba Bataknese society, the housing has a linear pattern, which created the hierarchy of space based on the familial pattern; while nature orientation became an important part that supported the position of the mass based on its functions, which are *jabu bolon* and *sopo*, and the direction of the housing development. There is also a space that is formed specifically as a public space called *alaman*. This became the orientation of the housing, where all kinds of community activities can be carried out. It is believed that there is a specific orientation in the development of Huta Raja housing which is to the east, and this orientation or direction is also followed by the physical development of the housing.

4.3. Physical characteristics

The physical characteristics of architecture in Huta Raja village is formed based on colors, textures, and ornaments, where the main ornaments are located in the middle of the building and become the center of attention, in the shape of ox head and *gorga*. The dominant colors that have become the colors of the Toba Bataknese tribe are red, white, and black, whereas the building development based on the directional orientation Eastward also strengthens the linear pattern of the housing.

4.4 .Activities of weaver community and the housing lay-out

There are two types of weaver community in Huta Raja village, namely the weavers who work in the housing area and those who work outside the housing area. The circulation of the community can be seen in Figure 5.

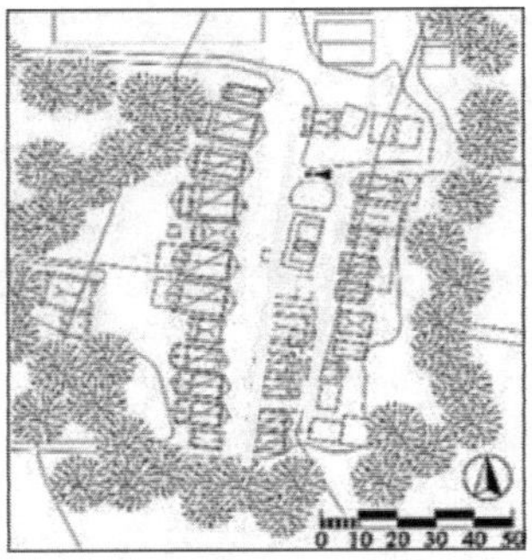

Figure 5. Circulation Pattern of Housing Community

4.5.Lay-out and ulos weaver community activities

The activity space of *ulos* weavers has been identified at the *alaman* area in the housing compound. This *alaman* became an informal production space to the community, and the activities are considered to be the common ones for the community. The activities of *ulos* weaver community are conducted mostly in the eastern area of *alaman*, therefore making the *alaman* becomes more public compared to the other areas in Huta Raja village. The eastern area of *alaman* tends to be busier and more crowded because it is used for the working space. The reasons why this area was chosen is because it has a larger space, and the weavers can be protected from the sun by the shade from the traditional buildings like *jabu-jabu*.

This shows that the characteristic of the eastern area are more private than the western area. Even though the difference is not directly limited by the space elements, it can still be seen clearly. The western area becomes the transitional space for the inhabitants of western part of Huta Raja village, from the private area to be a more public area. Even though the area has become more public, the western part of the village is bordered by housing mass, so the production zone is located inside the housing zone with the same circulation path, which is the *alaman*.

Weavers start their activities from their house to the *alaman* to weave. Weaving activities are carried out in front of several houses to protect the weavers from the sun. If the weather does not permit them to work outside (rain or overheat), the activities are moved indoors to their respective houses. Weaving activities are usually conducted from 9 a.m., followed by a communal lunch, and continued with further weaving until the afternoon. In the evening, usually there is a meeting to discuss events and activities related to the *ulos* sales. See Figure 6.

Figure 6. Activity pattern of weavers inside the housing compound
N.B. Hunian = Living Space; Alaman = Public Space

4.6.Lay-out of the outside area the housing compound

Based on the mapping of the activity pattern of the *ulos* weavers, it can be identified that the activity space formed at the housing compound is located in the *alaman.* This *alaman* became the informal production space for the community, because there are other functions besides production in the *alaman* space, such as the reception area, a common area for cultural and other events, the main circulation at the housing compound, a vehicle parking space, and an interactive area for the inhabitants as well as the visitors.

With those various functions of the *alaman,* the *ulos* production activities are included in the commercial category, albeit an informal one. If supported by the large number of weavers (as a community), visually the common weaving activity in one space will show an interesting diversity, especially for the tourists. The building mass in the eastern area is dominated by traditional buildings that are taller and bigger than the ones in the western area.The buildings in the western area are only dominated by *sopo, jabu, non-stage jabu,*and one-storey modern housings. See Figure 7.

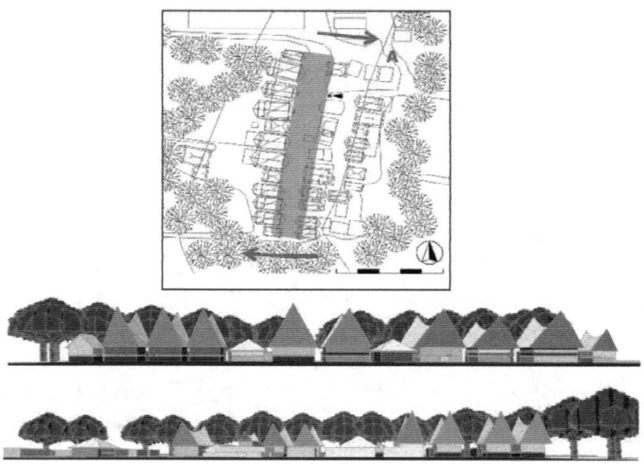

Figure 7. Building mass position in Huta Raja Village

Based on the space characteristics, the eastern area is more private than the western one. Even though the difference is not directly limited by the space elements, it can still be seen clearly. The western area becomes the transitional space for the inhabitants of the western part of Huta Raja village, from the private area to the more public area. Even though the area becomes more public, the western part of the village is bordered by housing mass, so the production zone is located inside the housing zone with the same circulation path, which is the *alaman*. The physical zone division can be seen in Figure 8.

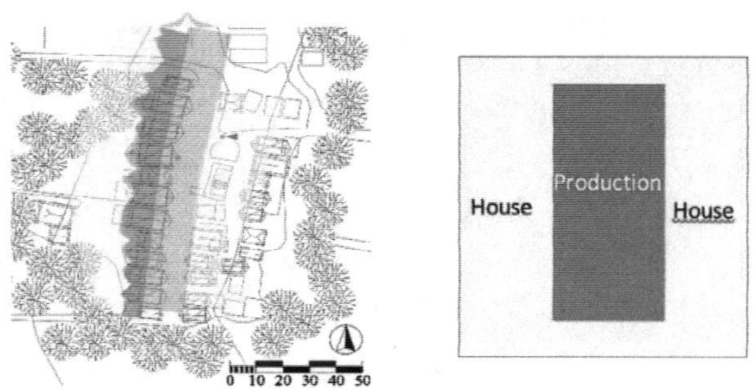

Red: housing
Yellow: Production
Orange: Transitional area

Figure 8. Physical zonation of Huta Raja Village

IOP Conf. Series: Earth and Environmental Science **903** (2021) 012007 doi:10.1088/1755-1315/903/1/012007

4.7. Indoor area lay-out

The indoor lay-out that is formed based on the activities of the *ulos* weaver community includes informal production space, where the physical space is formed based on the influence of building mass on the housing space. This condition can be seen in the two traditional building masses that were identified to seek the relations between the outdoor and indoor areas. These two buildings are the large *jabu* and the small *jabu*.

The indoor area in the large *jabu* owned by a weaver in Huta Raja village is categorized as the mixed space type, where the housing becomes one with the working space; and the main role of the building is as the living space. The weaving production took up the space that previously was used as a storage, which now has become more interactive with the outdoor area. See Figure 9.

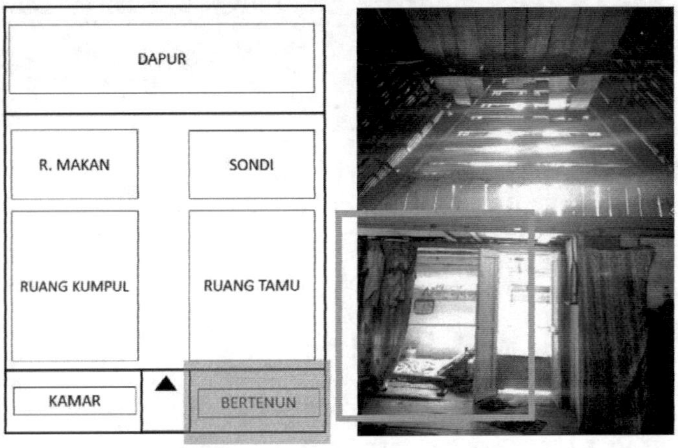

Figure 9. Weaving area in large *Jabu*
N.B. Dapur = Kitchen, R.Makan = Dining Room , Ruang Kumpul = Meeting Room , Ruang Tamu = Guest Room, Kamar = Room, Sondi = Bertenun = Weaving Room

In the small *jabu*, the composition of the space is like in Figure 10, where the weaving area is located at the front area of the building, that previously functioned as *jambur* room or terrace. The open space type makes it easier for communication between the weaving community and the other community in the *alaman*. Therefore, the small *jabu* is categorized as a mixed type, where the living area is embedded with the production area, and the main function of the house is providing living space.

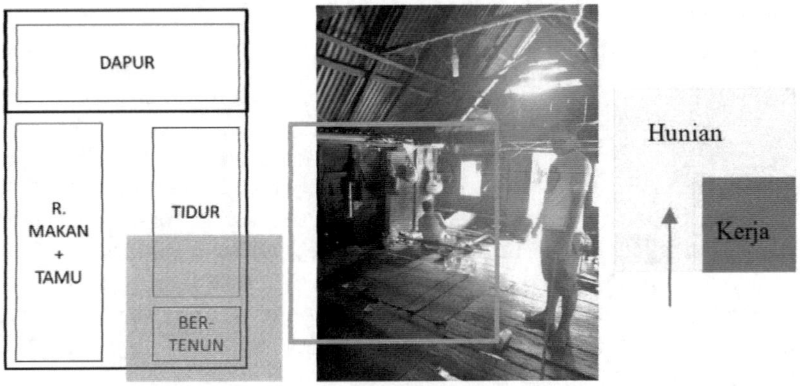

Figure 10. Weaving area in small *Jabu*
N.B. Dapur = Kitchen, R.Makan = Dining Room, Ruang Tamu = Guest Room, Kamar Tidur = Bedroom

IOP Conf. Series: Earth and Environmental Science **903** (2021) 012007 doi:10.1088/1755-1315/903/1/012007

Based on the description above, there are differences between the front area in the large *jabu* and the small *jabu* in providing the space for *ulos* weaving activities. See Figure 11. People who are allowed to weave in the area are the family or friends of the house owner. The relationship with the *alaman* can be seen in Figure 12.

Figure 11. The difference between weaving area in the small and large *Jabu*

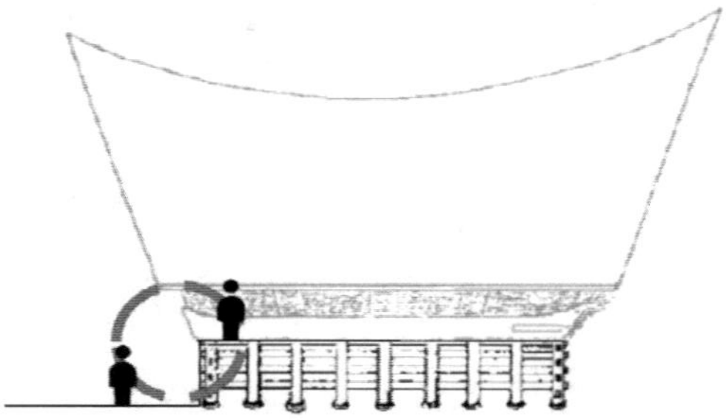

Figure 12. Relations between weaving area and *alaman*

Based on the identification results, the weaving production space inside the building took up the front area of the building that is modified into a more interactive one with the outdoor space by providing windows. The composition of space in the small *jabu* is similar with the large *jab*u, the only difference being the size of the middle area that is smaller than the large *jabu*. The modification is made by taking the space of the front area (that was supposed to be a *jambur)* and changing it into a weaving area by adding the windows to interact with other weavers in the *alaman*. There are no weavers working at the front of small *jabu* house, due to the small size of the *jabu* that is unable to protect the people from the sun or inclement weather. Therefore, the owner of the small *jabu* usually weaves alone inside his house, or joins the others in front of the large *jabu*.

5. Conclusion

The main factor that forms physical housing space is culture. Other factors, such as climate, geographical location and conditions, politics and economics are the modifying factors. The main factor has been identified with Huta Raja village as a manifestation. The geographical location determines the borders of housing space, so we can identify the geometric shape of Huta Raja village and its direction of development. The landscape of Samosir helps the direction and position of mass location in the Huta Raja housing compound related to Toba Bataknese cultural elements, primary needs of Lake Toba, and the flat land condition in the area.

The physical elements in the housing compound, in this case the fostered environment, facilities and infrastructure, do not affect the physical space form of Huta Raja village, but only affect the private characteristic of the housing compound. The linear pattern of the compound has formed the mass location according to the familial hierarchy of the Toba Bataknese tribe, and to the nature orientation. The area with specific type in Huta Raja village became the transitional area of the two housing lanes mentioned in the previous point.

Nature orientation system became the orientation of the development direction in Huta Raja village that had started from the *bolon* house, *sopo*, and other buildings. Colors, textures, and ornaments have formed the physical housing space with zonation based on the groups that have similar characteristics. The temperature in the housing area influenced the forming of outdoor groups, thus considerably enlivening the public space.

Huta Raja village area has a group of community with specific activities, in this case the *ulos* weaver community, whose members created the need of space to carry out the weaving activity. This need did not create a new space, but took over the space that has been combined and blended with the housing area for indoor activities, and used the *alaman* (existing space) to communicate between each other to discuss the issues of *ulos* production and sales.

The *Ulos* weaver community in Huta Raja village influenced changes in several function in the housing compound. There are some groups of space for weaving in *alaman* area, where the location follows the large building, which is the large *jabu*, especially to protect the community from the sun by sitting under the shade of the *jabu*, Therefore, the public area of *alaman* has changed from a total public area, to partly becoming a production area.

The changes also happened inside the houses in Huta Raja village. Every weaver changed their storage into a production area, and adapted their house into a mixed-type house, with an informal production area that is also used for other activities. The weaving room inside the building also adapted the mixed-type pattern because it is a part of the same mass with other living areas (bedrooms, living room, and others).

The community of *ulos* weavers that became a special characteristic of Toba Bataknese traditional housing in Huta Raja village should receive better facilities and infrastructure for their production space, both in the *alaman* area or in certain buildings, in order to make the weavers more comfortable and protected from the sun, but still visible to the public or visitors. The uneven spread of the working areas due to the shadows thrown by some buildings had created a chaotic outdoor area that made it seem unpleasant.

As a Batak Toba traditional housing compound, Huta Raja village has adapted a linear pattern of physical form, where the housings are facing each other and surround the common area. This shape needs to be preserved, because it has been determined by the rules laid down by and inherited from the ancestors. In the future, the development of Huta Raja village should also follow the same orders: nature orientation and linear pattern, so the cultural elements will always be preserved.

The shape and façade of traditional housing mass should also be preserved to maintain the housing characters. The building masses that have already been built in modern design can be located outside the two linear patterned compounds, so they will not destroy the cultural space in the *alaman*. The conservation of physical spatial form of Toba Bataknese housing compound can potentially be considered as cultural tourism destination, especially if supported by the existence of the *ulos* weaver community.

Culture, social structure, and physical order of traditional housing are the local elements that need to be conserved for maintaining the identity of local culture. The cultural diversity can be part of a tourist destination that will attract tourists and visitors from around the world.

IOP Conf. Series: Earth and Environmental Science **903** (2021) 012007 doi:10.1088/1755-1315/903/1/012007

6. References

[1] Hidayati, N. 2020. *Revitalisasi Tumah Adat di Kampung Ulos Samosir*. Portal Kementerian PUPR.hal : 20-25.

[2] Rapoport, A. 1969. *House Form and Culture.* London, Pearson.

[3] Depdikbud. 1997. *Arsitektur Tradisional Sumatera Utara.* Jakarta: Proyek Inventarisasi dan Dokumentasi Kebudayaan Daerah.

[4] Domenig, G. 2003. *Consequences of Functional Change Granaries. Granary- Dwellings, and Houses of Toba Batak. In Domenig, Indonesia Houses Tradition and Transf ormation in Vernacular Architecture.* Hal : 61-97 [7] Kusumawardhani, Ratu Arum. 2012. *Sejarah Perumahan di Jakarta*. Jakarta: Dinas Perumahan dan Gedung Pemda DKI Jakarta. hal: 3-4, 94-95.

[5] Tobing, R R, Sakti, A K, Hutabarat, G M. 2018. *Bentukan Arsitektural Perumahan dan Permukiman Tradisional berdasarkan Keberadaan Marga Suku Batak Toba di Desa Huta Ginjang Samosir.* Bandung: Universitas Katolik Parahyangan.

[6] Sasongko I, 2005. *Pembentukan Struktur Ruang Permukiman Berbasis Budaya.* Jurnal Dimensi Teknik Arsitektur 33 (1). hal : 1-8

[7] Crow, G, Allan, G. 1995, *Community Life.* Harvester Wheatsheaf.

[8] Habraken, NJ, 2000, *The Structure of the Ordinary.*The MIT Press

[9] Silas, J, 2000. *Rumah Produktif dalam Dimensi Tradisional dan Pemberdayaan.* Laboratorium Perumahan dan Permukiman ITS, UPT Penerbitan ITS.

[10] Tobing, RR, Siahaan, U, & Kridarso, ER, 2016. *Model Rumah Produktif Batik Keturunan Cina di Pekalongan.* Bandung: Universitas Katolik Parahyangan.

[11] Bungin, B, 2020. *Metode Penelitian Kualitatif,* LPPM Universitas Negeri Jember.

Privately Managed Green Open Space and Its Ecological Role in Improving Thermal Comfort of *Kesiman Kertalangu* Area

I G Edhy[1], N K A Siwalatri,[1] N M Yudantini,[3] and Gusti Ayu Made Suartika[1]

[1] Master Program in Architecture, Development & Planning of Udayana University - Bali, Indonesia

Email: gede.edhyputra@gmail.com

Abstract. The rapid development of urban areas in general has a negative impact on built environmental. The decrease of green open space is cause by increasing the need of space for human activity. The reduced area of green open space in Denpasar City, especially in private green open space in the *Kesiman Kertalangu* area, causes environmental quality to decline. Reducing the area of green open space can cause urban problems, such as increasing city air temperatures. Therefore, it is important to know the role of private green open space in residential yards for increasing thermal comfort in the *Kesiman Kertalangu* area. For this study the method used is a qualitative research method with a descriptive case study approach. The aim of this study is to determine the role of private green open space in residential yards in term of thermal comfort. The results show that in residential yards that have relatively large areas of land, the existing private green open space conditions are quite optimal. However, in residential yards that have limited land area, the availability of green open space is very minimal. The existence of private green open space in residential yards at the Kesiman Kertalangu area in general is not able to work optimally in reducing temperatures to create a suitable thermal comfort for the environment.

Keywords: private green open space, ecology, thermal comfort

1. Introduction

The sustainability in urban area is very important. This is due to the rapid development of urban areas which have a negative impact on built environmental such as happened in Denpasar City. The need of space for community activities and residences has reduces the availability of existing green open spaces and tends to decrease. Green open space is starting to be converted into built space, especially residential houses. This condition causes the role of private green open space in residential which is to support the existence of urban green open space (RTHK) as a whole, especially in carrying out its main function as an ecological function is decline.

The impact that is starting to be felt as a result of the reduced green open space as a result of the increasingly densely built space is the increasing temperature of the city in Denpasar, especially in the Kesiman Kertalangu area. The lack of green open space in this function is exacerbated because people tend to ignore the existence of green open spaces ecologically [1]. Such dynamics are feared in the future there will be a decrease in the ratio between built space and green open space in Denpasar City [2, 3]. This condition is quite a dilemma, especially for urban areas, because the area of residential land is relatively limited because of the high price of land [4, 5], so a strategy is needed to improve the quality of private green open space in carrying out its function as shading elements (reducing temperature)

The existence of a good and optimal green open space can actually function well in reducing temperature. Because one of the important functions of green open space is as a shading element that can improve the quality of the microclimate, especially the thermal comfort of the surrounding area. The way that can be done to maximize the function of private green open space in residential yards to reduce temperatures is to improve the quality of the existing green open space by improving the quality of the planted vegetation and maintaining a minimum of 10% green open space. The purpose of this study was to determine the role of private green open space in optimizing thermal comfort in the Kesiman Kertalangu area

2. Theoretical Frameworks

This study uses several theoretical foundations to determine the role of private green open space in residential yards in improving thermal comfort at Denpasar city especially at Kesiman Kertalangu area, including the theory of green open space, and private green open spaces, the ecological function of private green open spaces as a shade or reducing temperature and the physical characteristics of vegetation in green open spaces.

2.1. Private Green Open Space

Private green open space is green open space whose land is privately owned and managed by the community or private sector [6, 7]. According to the Regulation of the Permen PU No 05 Tahun 2008 concerning Guidelines for Provision and Utilization of Green Open Spaces in Urban Areas, private green open spaces are green open spaces owned by certain institutions or individuals whose use is for limited groups, including gardens, house yards, public buildings. and private public property planted with plants [16]. Then according to [8] in this book "Urban Green Space System Planning, Landscape Planning" private green open space is a green area such as a private garden on institutional garden that is managed and privately owned. From this understanding, it can be concluded that private green open space is a green area whose land is privately owned and managed by the community and private sector and its use is for limited groups. Based on the Permen PU No 05 Tahun 2008 the classification of private green open space consists of several types, namely green open space in residential yards, green open space for offices, shops, business premises and green open space in gardens and roofs of houses, whose land is privately owned. In this research, the context of private green open space is green open space whose land is privately owned and managed by the community and it is private green open space in residential yards.

2.2. Ecological Function Private Green Open Space as Shading Elements

Green open space is a very important element in shaping the structure of the city and maintaining the survival of the ecosystem in it. That is why green open space is very important and has the main function as an ecological function. Green open space performs its function ecologically, aiming to ensure the provision of green open space as part of the air circulation system, as a shade (reducing temperature), reduce noise, oxygen producer, reduce carbon dioxide, water catchment area, and as a windbreak [9].

The role of green open space in improving the quality of the environment ecologically. Humans cannot live alone without plants and animals around them, besides that humans also need air, light, water, soil, heat and certain climatic conditions to live. Between living things and their environment mutually influence each other [10]. A unified system between humans, plants, animals and their environment is called an ecological system. In this ecological system, in relation to the environment there are four important elements, namely air (wind), water, earth, fire (light). Green open space in this case has a role to balance this, one of which is to function as shading elements to reduce temperature. Green open space in which there are plants, grass and trees will have a very significant influence on the air temperature around it. The existence of green open space will be able to protect from radiation and heat from the sun which keeps the temperature and humidity of the air cool and comfortable. Shaded areas with green open spaces will receive less radiation and heat from the sun which makes the air temperature lower. According to [11], the ideal climatic conditions for humans are when the air is clean and the air temperature is between 27^0-28^0C. If you look at the average temperature in every big city in Indonesia, it is not easy to reach this state. For this reason, it is important to procure a green open space in accordance with existing provisions, theories and regulations, to support and optimize its function ecologically in maintaining the microclimate and temperature of the city.

To optimize green open space, especially in the private green open space of the residential yard in carrying out its function as a shade or reducing temperature, the most important thing is the availability of green open space that has an optimal area. In the arrangement of a city space, the provision of green open space in residential homes is regulated by the green basic coefficient (KDH). These rules vary by region. In Denpasar City, the minimum provision of green open space in residential yards is 10-28 percent of the total land area. Based on the theory from GBCI (Green Building Council Indonesia), every residential building must have a green basic area (RTH) in the form of vegetation (soft scape) that is free from buildings and a pavement of at least 10 percent of the total land area. This means, to optimize the function of green open space in reducing temperature, each residential yard must provide private green open space at least 10 percent of the total land area and free for buildings and pavement. If it is below this area, it can be determined that the existing green space will not be able to function optimally.

2.3. Vegetation in Green Open Space

The role, benefits and criteria of vegetation in green open spaces in improving environmental quality in urban areas are as shading, noise damper, carbon dioxide absorber, oxygen producer and as a windbreak [17, 19]. Vegetation has a important role in life. The existence of green open space is very dependent on the types of vegetation planted. Trees, shrubs and grasses can change the temperature of a city and reduce air pollution, and leaves on trees can control noise and break up sunlight. The effectiveness of the vegetation is highly dependent on the species, for example the characteristics of shady leaves, branching and many branches, root types, crown width and plant size [12].

The ability of a vegetation to reduce temperature, is dependent on the type and characteristics of the tree. Each type of tree has its own characteristics that make it have different effectiveness in carrying out ecological functions. Based on the ecological function of green open spaces as shade, the criteria for plants that have a wide crown [13]. Plants with a wide crown allow it to withstand radiation and heat from the sun, so that the air temperature around the green open space will be more comfort. Green open space in which there is vegetation with a wide crown will be able to reduce the air temperature during the day and vice versa at night will keep the air temperature warm. Its effectiveness is highly dependent on the characteristics and types of plants, the position of the sun's fall and weather conditions. Plants with physical characteristics of shady leaves and wide leaf surface areas are also good at reducing temperature. With shady leaves and a wide leaf surface area, the photosynthesis process can take place better, so that the amount of carbon dioxide from air pollution can be absorbed more and vice versa, the oxygen produced will be the same. This will make the air around the green open space cooler.

Figure 1. The Role of Trees in Reducing Temperature
Source: [13]

According to Laurie (1986), the characteristics of a tree will be very effective in controlling temperature if it has the physical characteristics of dense, shady and evergreen leaves. Trees with these characteristics will be able to control light, thus making the area receive less heat and sunlight which makes the air temperature under the tree's shadow lower. The trees with relatively high physical characteristics are also good at reducing temperatures [13]. Trees with these characteristics have a wider range of protection against heat and sunlight, as a result the area under and around the tree will be protected from radiation so that the air temperature around the tree will be maintained. From the explanation above, it can be concluded that the existence of trees in green open spaces is very important, especially in carrying out their function as shade, namely to reduce air temperature. The effectiveness of these functions is highly dependent on the type and physical characteristics of the tree. The physical characteristics of a tree that function well as shade are (1) a relatively tall tree, (2) having a wide tree crown, (3) the characteristics of shady and thick leaves (4) having a wide leaf surface area.

3. Research Methods

In this study, the method used is a qualitative research method with a descriptive case study approach, namely a process of collecting data and information in detail, intensively, holistically and systematically about a situation, to describe a situation or object in depth and detail in a limited scope [14].

The research location is in the city of Denpasar, namely in the village of Kesiman Kertalangu. The boundaries of the selected research area include two Banjars, namely Banjar Kertalangu and Banjar Kertapura. The boundaries of the research area are as follows: in the east it is bordered by Jl. By Pass Ngurah Rai, in the north it is bordered by Jl. WR. Supratman, in the west it is bordered by Tukad Ayung and in the south by Jl. By Pass Ngurah Rai. The selection of this area as the research boundary was due to the discovery of various cases showing that the availability of private green open space was not optimal, then the large number of research samples and various strata of residential houses, as well as the ease of obtaining data.

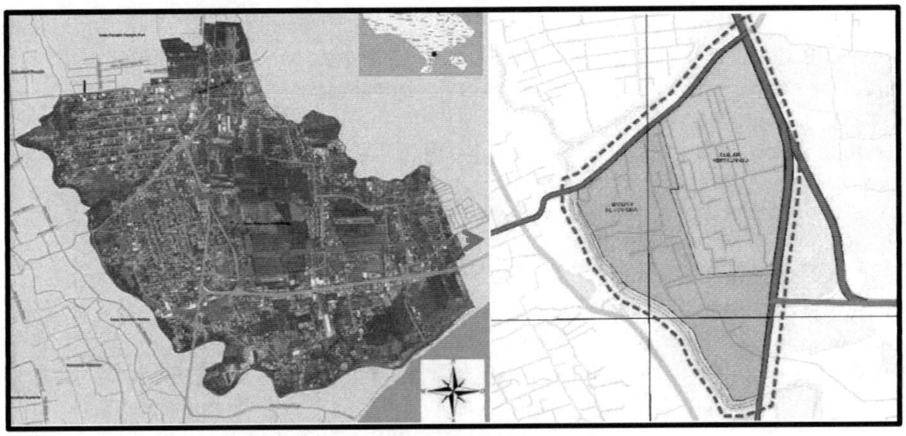

Figure 2. Research Area
Resource: [15]

In this study, the type of private green open space that will be studied is private green open space in residential yards. Green open space is further divided into 3 types of green open space in accordance with the Regulation of the Permen PU No 05 Tahun 2008 concerning Guidelines for Provision and Utilization of Green Open Space in Urban Areas, namely (1) private green open space at residential yards with small yards, (2) private green open space at residential yards with medium yards and (3) private green open space at residential yards with large yards. The three types of green open space are taken with different samples and will be selected randomly according to the criteria that have been determined for each strata.

Table 1. Research Sample

No	Types of Green Open Space	Number of Samples
1	Private green open space at residential yards with small yards (< 200m2)	25
2	Private green open space at residential yards with medium yards (≥200m2 and ≤500m2)	20
3	Private green open space at residential yards with large yards (> 500m2)	5

Table 2. Research Variables

Variable	Sub Variable	Assessment criteria
Private green open space area	Green area	Percentage of Green Area (KDH Denpasar City, GBCI (Green Building Council Indonesia)[18]
Vegetation	As Shade (Reducing Temperature)	Tree with wide crown [13] The leaves of the tree are shady and thick [11] Tree have a wide leaf surface area [13] Tres are tall [13]

IOP Conf. Series: Earth and Environmental Science **903** (2021) 012008 doi:10.1088/1755-1315/903/1/012008

To analyze the optimization of private green open space in carrying out ecological functions as shade or reducing temperature, two key variables are used, namely the area of private green open space and the physical characteristics of trees that are good at reducing temperature. Each variable has assessment criteria or parameters that become benchmarks in determining whether or not a green open space is effective in reducing temperature. In the following Table 2 describes the research variables.

To measure the variable area of private green open space, the assessment criteria used is the minimum area of green area, which is 10 percent of the land area. Private green space can be said to be optimal in carrying out its ecological function as a shade (reducing temperature) if it meets these requirements. Then the vegetation as shade will be assessed based on the physical characteristics of the tree using a rating scale/score for each tree in the Private Green Open Space in residential yards at Kesiman Kertalangu Area. Until later will be obtained tree species that have good characteristics in carrying out the function as a shade (reducing temperature). From the results of the assessment of these variables, the data analysis process that will be carried out qualitative research method with a descriptive case study approach. The results of the research will be compared with the theory and observations in the field on the area of green open space and the physical characteristics of trees in reducing temperature. Until later it will be obtained how optimal the role of private green open space in the residential yard at Kesiman Kertalangu Area is in reducing temperature to increase thermal comfort in the surrounding environment.

4. The Role of Privately Owned Green Open Space in Improving Thermal Comfort

Analysis of the function of private green open space in increasing thermal comfort to find out how effective the current private green open space in the Kesiman Kertalangu area can affect the temperature conditions in the environment so that it can form a good thermal comfort for the environment and the community. Therefore, the analysis is carried out on two important indicators, namely an analysis based on the area of private green open space and an analysis based on the existing vegetation in private green open space.

4.1. Analysis Based on Area

The area of private green open space in residential yards in the Kesiman Kertalangu area shows different results for each residential strata. Overall, the results of the study indicate that there is still a need to optimize the provision of private green open space, especially in residential yards that have a small and limited land area. If you look at the theory, in order to maximize its function as a shade, namely to reduce the temperature, private green open spaces must at least have a Green Area (DH) of at least 10 percent of the total land area. The 10 percent is green open space in the form of vegetation (softscape) both plants, grass and trees that are free from buildings and sidewalks (Green Building Council Indonesia). The results of observations of 50 samples of private green open space for residential yards in the Kesiman Kertalangu area as shown in Figure 3, show the results that there are 16 percent of residential yards that have private green open space area of more than 28 percent of the total land area. Then there are 38 percent of residential yards that have private green open space between 10-28 percent of the total land area. And there are still 46 percent of residential yards that have private green open space below 10 percent of the total land area. This means that there are still many residential yards that have private green open spaces that are not optimal. As a result, existing green open spaces at residential yard will not be able to function optimally in lowering temperatures and creating thermal comfort for the surrounding environment.

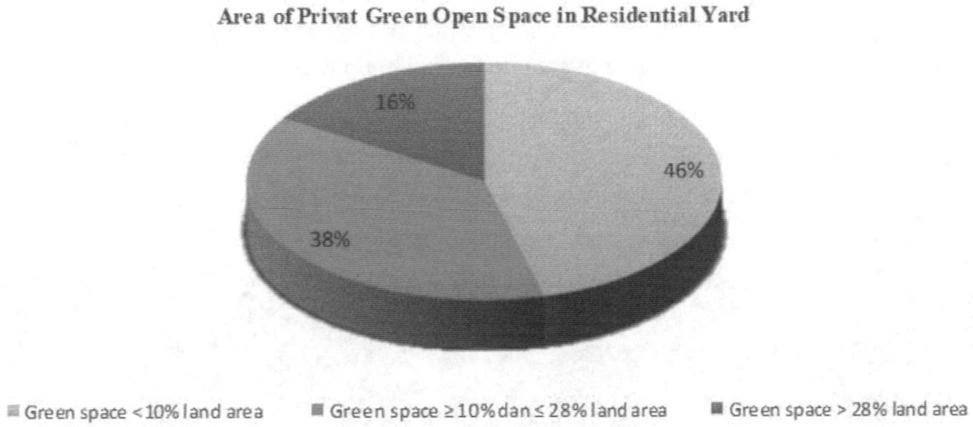

Figure 3. Percentage of Private Green Open Space in Residential Yards

The use of private green open space in the residential yard at Kesiman Kertalangu area has largely received less attention from the public. Private green space is seen more aesthetically and economically, rather than its main function: the ecological function. Private green open space to reduce temperature has not functioned optimally in residential yards with relatively small land areas. This is because most of the land is built space and pavement, both made of concrete, paving stones, brushed coral, stone, and ceramics. So only small amount land is green space, planted by grass, shrubs or trees. Land that is dominated by pavement will absorb heat quickly through radiation and sunlight which causes the air temperature around it to be increase. Unlike the case with land which is dominated by plants, it allows the area to form a protection zone against radiation and sunlight, forming a shadow that makes the air temperature around it cooler so that thermal comfort can be achieved. Figure 4 below shows the condition of private green open space on relatively limited land.

Figure 4. The condition of private green open space in the residential with a limited area

In residential yards that have wider land, the provision of green open space is more optimal, the existing land is also more dominated by green open space. Observation data shows that in residential yards that have a wider area, it is use more for private green open space. The existing open space is dominated by green open space of grass, shrubs and trees that allow the area to work optimally in lowering the temperature. Plants and trees, the air temperature around private green open spaces will be lower, this is because plants and trees can protect from radiation and solar heat, trees can create a protected zone so that the area below has a lower air temperature. In addition, the existence of optimal green open space will allow trees or plants in the area to reduce air pollution, absorb more carbon dioxide

IOP Conf. Series: Earth and Environmental Science **903** (2021) 012008 doi:10.1088/1755-1315/903/1/012008

and vice versa, produce the same amount of oxygen. So that the air temperature in the area around the private green open space becomes cooler. Figure 5 shows the condition of private green open space on the yard which has a wider area with the availability of more optimal private green open space

Figure 5. The condition of private green open space in the residential yard with a large area

4.2. Analysis Based on Vegetation

The types of vegetation found in the private green open space in the residential yards at the Kesiman Kertalangu area are quite diverse and have relatively large numbers, especially in the yards of residential houses that have large areas of land. Of the 50 samples of residential yards that have been observed, there are 20 types of vegetation in the private green open space of residential yards including *Pseudosasa Japonica, Averrhoa Blimbi, Araucaria Heteropylla, Magnolia Alba, Erythrina Cristagali, Delonix Regia, Psidium Guajava, Plumeria Rubra and Obtusa, Cocos Nucifera, Terminalia Mantaly, Bougainvillea Spectabilis, Mangifera Indica, Artocarpus heterophyllus, Morinda citrifolia, Pandanus Tectorius, Veitchia Merillii, Hibiscus Tiliaceus, Ptychosperma Macarthurii, Zyzygium Aqueum*. The most dominant vegetation in private green open space is *Plumeria Rubra*. This type of vegetation is planted almost in all residential yard, both yards of small, medium and large areas. The sizes and heights also vary. The private green open space in the residential yard, there are also several types of trees with relatively large sizes and wide crown, such as *Mangifera Indica, Magnolia Alba, Morinda Citrifolia, Terminalia Mantaly and Hibiscus Tiliaceus*. However, this type of tree is only often found in residential yards that have wider land. This is because the large area of land allows the provision of green open space of the same extent, so that a more diverse use of vegetation can be chosen with a larger size and wide tree crown.

Overall, the condition of the vegetation in the private green open space of the residential yard at the Kesiman Kertalangu area is relatively good however the selection of vegetation does not take into account the ecological function, especially its ability as a shade in reducing temperature. The choice of vegetation takes into account are aesthetic and economic functions, so the tree species that are often found are flowering and fruit trees such as *Plumeria Rubra, Psidium Guajava, Cocos Nucifera* and *Bougainvillea Spectabilis*. The choice is also motivated by the culture of the Balinese people who use a lot of flowers for their offerings.

To measure whether each tree is effective in reducing temperature, a rating scale or score is used with the parameters of the physical characteristics of the tree: the tree with wide crown [13]; the leaves of the tree are shady and thick [11]; trees have a wide leaf surface area [13]; and trees are tall [13].

Table 3. Criteria for Assessment of Vegetation as Shade (Temperature Reducing)

No	Type of vegetation	Criteria as a shade				Score (%)	Category
		K1	K2	K3	K4		
1	Pseudosasa Japonica	2	2	1	2	43,75	KB
2	Averrhoa Blimbi	3	2	2	3	62.5	BA
3	Araucaria Heteropylla	2	3	2	4	68,75	BA
4	Magnolia Alba	3	4	4	4	93,75	SB
5	Erythrina Cristagali	4	4	3	4	93,75	SB
6	Delonix Regia	4	4	3	4	93,75	SB
7	Psidium Guajava	3	3	3	3	75	BA
8	Plumeria Rubra	3	3	4	3	81,25	SB
9	Plumeria Obtusa	3	3	4	3	81,25	SB
10	Cocos Nucifera	2	1	2	3	50	KB
11	Terminalia Mantali	4	3	3	4	87,5	SB
12	Bougainvillea Spectabilis	3	3	2	2	62,5	BA
13	Mangifera Indica	4	4	4	4	100	SB
14	Morinda Citrifolia	2	3	2	2	56,25	KB
15	Artocarpus heterophyllus	3	3	4	4	87,5	SB
16	Pandanus Tectorius	2	2	2	2	50	KB
17	Veitchia Merillii	2	1	2	2	43,75	KB
18	Hibiscus Tiliaceus	4	4	4	4	100	SB
19	Ptychosperma Macarthurii	2	2	2	2	50	KB
20	Zyzygium Aqueum	3	2	2	2	56,25	KB

Note : K1 : Tree with wide crown
K2 : The leaves of the tree are shady and thick
K3 : Tree have a wide leaf surface area
K4 : Tree are tall

Rating:

Score 4 : Very good (SB) if ≥ 81% criteria are met
Score 3 : Good (BA) if 61-80% criteria are met
Score 2 : Not good (KB) if 41-60% criteria are met
Score 1 : Bad (BU) if ≤ 40% criteria are met

The results of the analysis show that the *Magnolia Alba, Erythrina Cristagali, Delonix Regia, Plumeria Rubra, Terminalia Mantali, Mangifera Indica, Artocarpus heterophyllus,and Hibiscus Tiliaceus* trees have very good physical tree criteria in reducing temperature or as shade. There are also types of *Averrhoa Blimbi, Araucaria Heteropylla, Psidium Guajava and Bougainvillea Spectabilis* which also have good physical characteristics in reducing temperature. If you look at each private green open space in the residential yard, this type of tree is not always there, except for the *Plumeria Rubra* tree which is almost in all private green open space in the residential yard. This shows that every residential expert has provided at least one type of tree that functions both as shade and can reduce temperature. Looking in more detail each type of tree has different physical characteristics so that its effectiveness in reducing temperature is not always the same. For example, if you compare each sample of residential yards that have a land area of less than 200m2 and an area of private green open space of less than 10 percent of the land area. The only tree species available are *Plumeria Rubra* trees, which in an analytical assessment have very good physical characteristics of trees in reducing temperature. The ability of the *Plumeria Rubra* tree in private green open space in reducing temperatures, especially during the day, is not too different where the temperature around the green open space at 11 o'clock in the afternoon shows 30^0 C. However, when compared to other samples, which have private green open spaces that reach more than 10 percent of the land area, and there are more diverse trees, namely the *Plumeria Rubra, Mangifera Indic*a and *Magnolia Alba* trees, the temperature around the private green open space during the day is more shady, showing a figure of 29.5^0 C. Physically, *Magnolia Alba* and *Mangifera Indica* trees have wider crown characteristics, shady and thicker leaves than *Plumeria Rubra* trees, so that the physical characteristics of these trees function more optimally ecologically in reducing temperature and creating thermal comfort in the environment. This is as said by Gray and Deneke (1978) that trees that have a wide crown will be better able to reduce the temperature of the air around them.

In other examples of cases in the private green open space, the residential yard which has a wider area of land and the area of private green open space, where the types of trees that exist are the type of *Plumeria Rubra* and *Mangifera Indica*. After measurements were made, the temperature around the green open space showed 29.4^0 C. When compared again with the residential yard that has wider land and private green open space, namely in the yard with a land area of more than 500m2, as well as more diverse tree species such as *Mangifera Indica* tree, *Plumeria Rubra, Magnolia Alba, Pseudosasa Japonica, Araucaria Heteropylla, Averrhoa Blimbi* and *Veitchia Merillii*. The results of measurements using an anemometer show that the temperature around the private green open space is much shadier at 29^0 C. This shows that the characteristics and variety of tree species are also able to affect the ability of private green open spaces in carrying out ecological functions as shade to reduce temperature, as stated by [12]. If it compared to the standard climatic conditions and ideal temperatures for humans, thermal comfort can be achieved if the air temperature shows the number 27-28^{0}C [11]. The climate and temperature conditions at the Kesiman Kertalangu area are still far from optimal to achieve this level of comfort. Therefore, it is necessary to optimize the current private green open space, both in terms of area and vegetation

From the analysis above, it is found that broadly, the availability of private green open space at the Kesiman Kertalangu area is still not optimal and has not contributed optimally to reduce temperature. Significantly, the existence of private green open space in the residential yard has not been able to create thermal comfort for the environment and the people in it. However, based on the vegetation analysis carried out, every house in the Kesiman Kertalangu area already has at least one type of tree that can function well in reducing temperatures, namely the *Plumeria Rubra*. This type of tree is found in almost every yard of a house at the Kesiman Kertalangu area. However, its effectiveness needs to be improved. The way is to plant new tree that are physically more capable in reducing temperatures. Such as the types of *Mangifera Indica, Magnolia Alba, Hibiscus Tiliaceus, Terminalia Mantali* and *Artocarpus heterophyllus*. In addition, the number of trees also affects, so planting many trees even though they have the same type in private green open spaces will further optimize the function of private green open space in reducing temperatures. Another way that can be done is to add gardens on the roof of the building or gardens on the fence walls and buildings (green wall). This method does not increase the area of private green open space, but its existence can function effectively ecologically to reduce the environmental temperature on a micro basis, absorb CO^2 and produce O^2 (Joga & Ismaun, 2011). This will certainly increase the function of private green open space in reducing temperature and will directly increase the role of private green open space in residential yards to create thermal comfort in the environment.

5. Conclusion

Based on the analysis that has been carried out, it is concluded that the role of private green open space in the residential yard at the Kesiman Kertalangu area in increasing the thermal comfort of the surrounding area is still not able to function optimally. Significantly its effectiveness can still be improved. If you look at it in more detail, its function is not optimal because in the residential yard that has a small and limited land area, the availability of green open space is very small, less than 10 percent of the land area. Likewise, there is not much vegetation. Most often found is a *Plumeria Rubra* tree which is almost in every residential yard. It is different for residential yard that have relatively wider land. The availability of green open space is more than 10 percent of the land area and the existing vegetation is much more diverse. However, this condition is not often found in residential yards at the Kesiman Kertalangu area.

Overall, the current private green open space in residential yard at Kesiman Kertalangu Area has been able to function to reduce temperature but its effectiveness is still not optimal. Optimization of the role of private green open space in increasing thermal comfort, namely reducing temperature, can be done by improving the quality of existing green open spaces by creating gardens on roofs, fences and walls of buildings. This method does not increase the area of private green open space but is very effective in reducing the environmental temperature. To increase the area of private green open space by replacing some of the pavement with plants, grass, trees or shrubs. Improving the quality of green open space can also be done by planting new trees, without reducing the existing trees. The selection of tree species must be more diverse and parked at a tight distance so that its function in creating thermal comfort in the environment can be optimal. Trees that can be selected are trees that have good physical characteristics in reducing temperature. Trees with wide crown, shady and thick leaves, wide leaf surface area, and relatively tall tree sizes such as. *Mangifera Indica, Magnolia Alba, Hibiscus Tiliaceus, Terminalia Mantali, Averrhoa Blimbi, Plumeria Rubra, Erythrina Cristagali* and *Artocarpus heterophyllus*.

6. Reference

[1] Groos and Lane 2007 *Landscapes of the Lifespan Exploring Accounts of Own Gardens and Gardening* (Journal of Environmental Psychology) 27(3) 225-241.

[2] Salain, Putu Rumawan. 2016 *Persoalan Tata Ruang Kota Denpasar Mendatang* (Denpasar)

[3] Suartika, G A M (2013) 'Lost in Translation: Balinese Vernacular Open Space' in Suartika, GAM (Ed) *Vernacular Transformations: Architecture, Place, and Tradition* Denpasar (Bali): Pustaka Larasan in Conjunction with Udayana University's Masters Program in Planning and Development for Urban and Rural Areas; Conservation of the Built Environment; and Ethnic Architecture.

[4] Suartika, G A M (2013) 'Open Space Traditions, Development and Modernity: The Case of Bali' *Journal Nasional Terakreditasi Mudra: Journal Seni Budaya* Vol. 28, No. 2 December 2013, Institut, Seni Indonesia Denpasar, UPT Penerbit, p: 283-299.

[5] Suartika, G A M, Mudra, I K, and Saputra, K E (2019) An Open Market: The Legitimation and Regulation of the Public Realm in Denpasar, in IOP Conf. Ser.: Earth Environ. Sci. 396 012041, p: 1-8.

[6] Joga, Nirwono dan Iwan Ismaun 2011 *RTH 30 % Resolusi Kota Hijau* (Jakarta : PT. Gramedia Pustaka Utama)

[7] Suartika, G A M (2019), 'Lessons from a Small Island – Density, Spatial Development, and Identity' in IOP Conf. Series: Earth and Environmental Science, 2019 IOP Conf. Ser.: Earth Environ. Sci. 248 012016, p: 1-13.

[8] Bilgli, BC and Gokyer E 2012 Urban Green Space System Planning, Landscape Planning (Croatia: InTech Janeza Trdine pp 107-112

[9] Nuraini, Cut 2009 *Peran Fungsi dan Manfaat Pekarangan sebagai Salah Satu Model Ruang Terbuka Hijau di Lingkungan Permukiman Padat Kota, Studi Kasus : Pekarangan di Karang Kajen, Yogyakarta* (Denpasar. Seminar Nasional)
[10] Frick, Heinz dan FX Bambang Suskiyanto 2007 *Dasar-Dasar Arsitektur Ekologis* (Jakarta: PT Kanisius)
[11] Laurie, M 1986 *An Introduction to Landscape Architecture* (New York: American Elsevier Publ.Co.Inc)
[12] Heston, Yudha P dan Dimas Hastama Nugraha 2017 *Oase di Tengah Kota : Kota Ekologis dan Penyiapan RTH* (Yogyakarta: Gadjah Mada Univesity Press)
[13] Grey, G.W. and Deneke, F.J. 1978 *Urban Foresty* (New York: John Willey and Sons)
[14] Yusuf, Muri 2014 *Metode Penelitian : Kuantitatif, Kualitatif & Penelitian Gabungan* (Jakarta: PT. Gramedia Pustaka Utama)
[15] Badan Perencanaan Pembangunan Daerah Pemerintah Kota Denpasar 2017 *Laporan Data dan Analisa "Identifikasi RTH 4 Kecamatan di Kota Denpasar"* (Denpasar)
[16] Departemen Pekerjaan Umum 2008 Permen PU No: 05/PRT/M/2008 Tentang *Pedoman Penyediaan dan Pemanfaatan Ruang Terbuka Hijau di Kawasan Perkotaan.*
[17] Ernawati, Rita 2015 *Optimalisasi Fungsi Ekologis Ruang Terbuka Hijau Publik di Kota Surabaya* (Surabaya: EMARA Indonesian Journal of Architecture)Vol 1 No 2
[18] Green Building Council Indonesia 2015 *Greenship Rating Tools untuk Kawasan* (Direktorat Pengembangan Perangkat Penilaian GBCI)
[19] Iskandar, Johan dan Budiawati S. Iskandar 2016 *Arsitektur Tumbuhan: Struktur Pekarangan Perdesaan dan Ruang Terbuka Hijau Perkotaan* (Yogyakarta: Teknosain)

Acknowledgement

Authors gratitude to the University of Udayana-Bali, Indonesia for enabling the publication of this article by providing research grants (with a contract No of: B/96-111/UN14.4.A/PT.01.05/2021, dated 3 Mei 2021).

The Existence of Stilt Houses Post-Disaster

Andi Hildayanti[1] and Wasilah[2]

[1] Field of Housing and Settlement Design, Department of Architectural Engineering, Universitas Islam Negeri Alauddin Makassar, Indonesia. Email: andi.hildayanti@uin-alauddin.ac.id
[2] Field of Architecture Design, Department of Architectural Engineering, Universitas Islam Negeri Alauddin Makassar, Indonesia. Email: wasilah@uin-alauddin.ac.id

Abstract. Stilt houses are one of the architectural identities of the archipelago that are still commonly found in Indonesia. Stilt houses have proven to be sustainable products because they still exist today after going through various natural disasters. This is inseparable from the stilts house building techniques, including the process of finding and determining the form and implementation of adat (philosophy). So, this research aims to show the existence of post-disaster stilt houses which include tsunami disasters in Aceh and Palu. This research is a case study that uses a comparativeanalysis method to show the influence factors of the existence of stilts houses after disaster in Aceh dan Palu. Based on the results of the study, it is known that the stilt houses in Aceh and Palu have stilt buildings that are sturdy and stable in accepting the force load when a disaster occurs. The key of building sturdiness lies on building structures such as connection systems, foundation type, material type, and the buildings proportion. These factors caused the stilts houses to not suffer severe damage, and in some cases did not show any damage to the buildings after being affected by natural disasters.

Keywords: Stilt house; disaster; building structures; Aceh; Palu

1. Introduction

Indonesia is an area located on the Pacific Ring of Fire, which is an area with a lot of tectonic activity. So, Indonesia continues to face the risk of volcanic eruptions, earthquakes, floods, and tsunami. In several incidents during the last 20 years, Indonesia made headlines in the world because of its terrible natural disasters that have caused millions of people and animals were dead, as well as destroying its land area, including a lot of infrastructures, and resulting in economic losses. Natural disasters are the most terrible events on earth because they cannot be predicted. Natural disasters can have a major effect on human life, can destroy the infrastructure of a city, and claim many lives, such as the Aceh earthquake and tsunami in 2004 and Palu in 2018. The location of Indonesia is at the junction of three earth plates that are constantly pushing and rubbing against each other. The more plates can be complicated and dangerous the fault is. This fault is the source of the earthquake. Earthquake energy from faults is not as big as earthquakes generated by collisions between plates. However, the dangers cannot be taken lightly because many of the faults are located right underneath the city, under people's houses.

Facing this situation, the archipelago people have long been trying to mitigate disasters by building earthquake-resistant traditional houses. One of the major earthquake events that occurred was in Minahasa in 1845 which damaged thousands of houses. After this incident, the Minahasa people started building houses with designs that minimize damage in the event of a similar disaster. Since this incident, people's houses have been built in small sizes, the poles are shortened and reduced and the frames of the houses are made in such a way so that they do not collapse easily.

The concept of a stilt house will be helping release earthquake energy so it doesn't hit the floor of the building directly. Damage and safety threats to residents can be smaller. This underlies the implementation of this research, to identify and recognize the concept of a stilt house and to know how it can survive and remain strong until now from the attacks of natural disasters, especially in the major tsunami disasters that occurred in Aceh and Palu.

2. Method

This research using the case study method, which is a research strategy, an empirical study that investigates a phenomenon in a real-life setting. The study locations chosen were the post-tsunami areas in Aceh and Palu. Research data were collected from various sources such as relevant previous research studies and the results of their observations. The data collected from the results of the investigation at the two locations were then compared to get conclusions about the concept of a stilt house (the criteria for the robustness of a building) that could withstand natural disasters. The variables studied were the stilts house which included the construction system, materials, and dimensions.

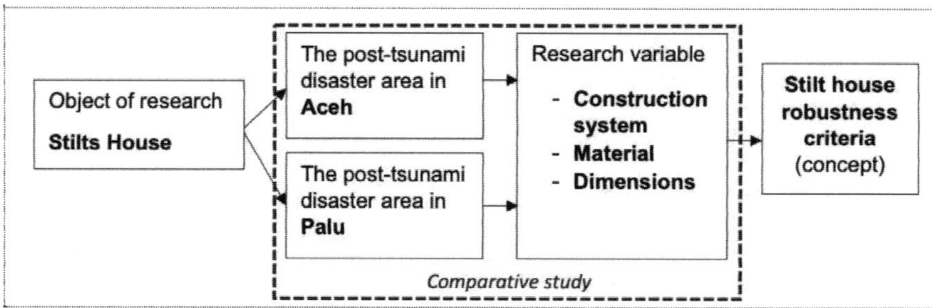

Figure 1. Research framework

Figure 2. Research location. *Source: Tribunnews, 2018*

3. Result and discussion

The results of this study explain the condition of traditional buildings (adat) that survive after natural disasters. The description is done separately.

3.1. The existance of stilt house in Aceh

The tsunami occurred on 26 December 2004 in Aceh. The tsunami triggered by an earthquake measuring 9.3 SR occurred at a depth of 30 km below the seabed and 100 km from the west coast of Aceh. The earthquake caused a tsunami wave as high as 9 meters which then swept across Aceh. This disaster killed more than 220 thousand people in Indonesia. This incident also caused casualties to areas in Peninsular Malaysia, Thailand, India, Sri Lanka, and the East coast of Africa. The Aceh Earthquake and Tsunami was the most devastating natural disaster with the largest number of casualties in Indonesian history. Now, Aceh has risen and built several places to remember this incident and provide lessons to the community.

Aceh is one of the areas most affected by earthquakes. Efforts to reduce the impact of disaster risk have been carried out by the community in ancient times. One of them is by building wooden houses that are not easily collapsed. One of the buildings that have survived is the type of stilts house. Where the roofing material is made of thatch leaves (but currently many are using zinc as roof coverings - see Figure 3) and several poles support the existence of the house. In addition, there are stairs to enter the house which are placed under the front.

Figure 3. Stilts houses survived from tsunami in Aceh. *Source: Husein and wicaksono, 2020*

The process of making this house-shaped on stilts entirely uses woods. In terms of construction, the placement of house poles led to the division of Aceh house space generally consisting of three pillar rooms, 16 or five pillar rooms 24. The wood used to make the house uses seumantok wood, the best type of wood in Aceh. This is one reason Aceh houses could last for hundreds of years. on the other hand, the construction is intertwined, the pillars of ordinary houses are made of sturdy wood, while thefloors and pillars of Acehnese houses are also of wood. To tighten the wood, the Acehnese ancestors did not use nails, iron, or concrete. But the wood is hollow, chiseled, and then given a hook. This interlocking technique has its purpose. One of them is to reduce vibration. By using a construction likethis, Aceh's house is resistant to shaking and does not collapse easily due to earthquakes.

Earthquake disaster mitigation efforts have been applied for a long time in Aceh. However, along with the times, there are changes in lifestyle and needs so that most residents have moved and adapted to modern times by building permanent houses (stone houses). However, not a few of them were aware of the natural conditions in Aceh which was prone and surrounded by an earthquake that had the potential for a tsunami, so they continued to build and maintain wooden houses or stilts houses.

IOP Conf. Series: Earth and Environmental Science **903** (2021) 012009 doi:10.1088/1755-1315/903/1/012009

Figure 4. Aceh traditional house shaped on stilts

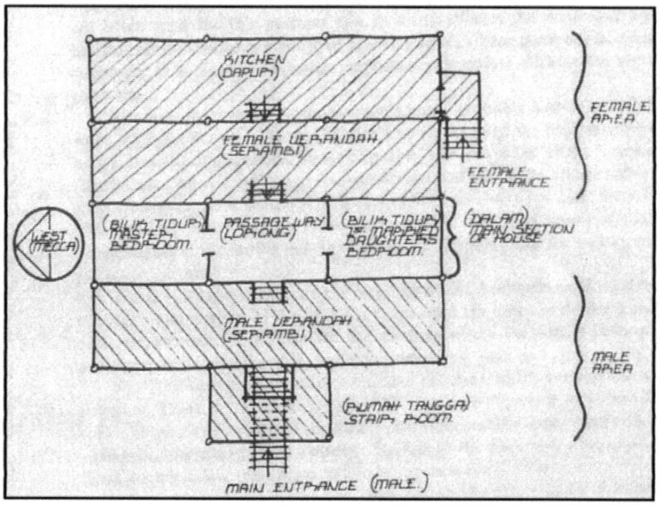

Figure 5. space function

Rumoh Aceh is one of tradisional Acehnese houses which has several constituent elements. Each of the elements that make up the Rumoh Aceh also has a distinctive name and philosophical meaning that is no less interesting. Here are some elements of traditional Acehnese houses and their philosophical meanings:

1. *Tameh*: a pole used as a support for the body of the house. This element is motivated by a typical Acehnese proverb, *"Kreuh beu beutoi kreuh, beulagee kreuh kayee jeut keu tamèh rumoh; Leumoh beu beutoi leumoh, beulagee taloe seunikat bubông rumoh"* (If it is hard, it must be as hard as the wood supporting the house; if it is flexible, it must be as flexible as the ropes for the roof of the house). The Acehnese philosophy of life in this regard is to be firm, but still soft-hearted.
2. *Tameh raja:* the king's pillar, the main pillar on the right side of the entrance. Called the king's pillar because of its size which is larger than some of the ordinary constituent pillars used.
3. *Tameh putroe:* the princess pole, which is the main pillar on the left side of the entrance. The princess pole is a pair of the king's pole. Called the pillar of the princess because of its position side by side with the pillar of the king
4. *Gaki tameh*: pillar legs, i.e. baseboards, usually of river stone. This pedestal serves to support the wooden posts so they don't sink into the ground.
5. *Rok*: ordinary locking beam. Its nature is to strengthen the relationship between the ends of each beam.

IOP Conf. Series: Earth and Environmental Science **903** (2021) 012009 doi:10.1088/1755-1315/903/1/012009

6. *Thoi*: locking beam that is perpendicular to the *Rok*.
7. *Peulangan*: where the inner wall (interior) rests.
8. *Kindang*: the place where the outer (exterior) wall rests.
9. *Aleue*: floor, made of small slatted boards.
10. *Rante aleue*: floor fasteners usually made of rattan or rope.
11. *Lhue*: truss beam for floor support.
12. *Neudhuek lhue*: the place where lhue rests.
13. *Binteh*: wall.
14. *Binteh cato*: chess wall, a form of wall braid.
15. *Boh pisang*: a small board on top of a kindang.
16. *Tingkap*: window. The windows of Aceh houses are made small. The main window is on the side of the house.
17. *Pinto*: door.
18. *Rungka*: roof truss.
19. *Tuleueng rhueng*: wuwung beam, where the rafters rest on the top end. These beams are made of light wood so as not to burden the roof.
20. *Gaseue gantong*: the legs of the horses.
21. *Puteng tameh*: the end of the chiseled pole, as a beam connector.
22. *Taloe pawai*: roof ties tied to the end of the *Bui teungeut*.
23. *Bui teungeut*: pieces of wood to hold the *neudhuek gaseue*.
24. *Tulak angen*: reject the wind, the cavity where the wind passes on the side wall of a triangular house

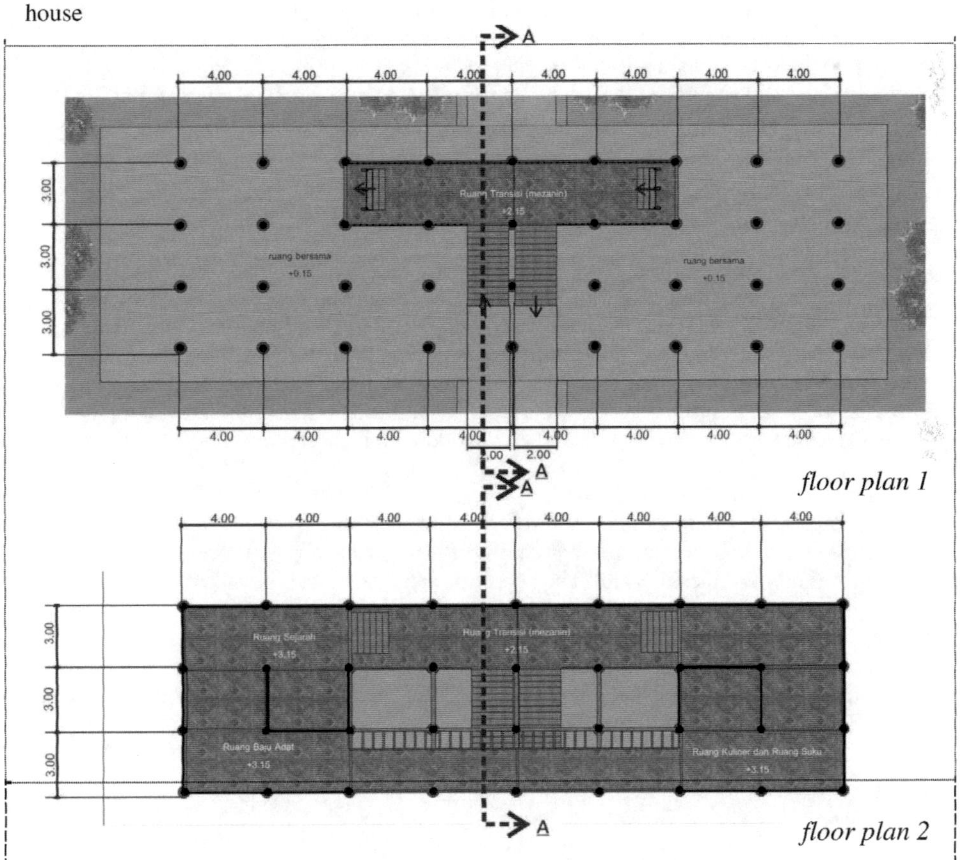

Figure 6. Floor plan of Rumoh Aceh

The open pillars of Aceh's houses have an important role in flood and tsunami disaster mitigation. This is because water can escape without any obstacles, thus Rumoh Aceh is safe from the threat of disaster. Several locations after the earthquake and tsunami disaster have also shown the robustness of traditional Aceh house buildings that have survived the threat of this disaster. Based on previous research studies, to prove Aceh's houses are earthquake resistant, laboratory tests were carried out through small miniatures and calculations of the SAP 2000 program. Widosari (2010) in his scientific journal "Maintaining Local Wisdom Rumoh Aceh in the Dynamics of Community Life After the Earthquake and Tsunami". reviews the toughness of the Aceh house. Based on these trials, the results obtained were that the Aceh house was proven to be able to withstand earthquakes because the main structure was strong and elastic. The key to this robustness and elasticity lies in the relationship between the main structures which interlock, only with pins, without nails, and form a rigid three-dimensional. Elasticity causes the building structure not to break easily, but only tossed from side to side and then back upright or liquefied (lifted) then able to back to its original place. When an earthquake occurs, even if the building moves, it is only about a few centimeters and is intact. A solid stone foundation that is only planted a little by five centimeters also flexes the movement of the entire building according to the movement of the ground.

So it can be concluded that three main structural components are central to the robustness of the building including the foundation (leg component) as the largest center of the building's load, then the poles and beams between the poles (body components) as load carriers from above and from the side, and the roof frame (component head) as a load support element at the top of the building and from the top side. There are at least two things that make Aceh's house strong and safe from disasters. First the poles, ornaments/carvings, and roofs. When the tsunami hit Aceh, the flood could pass through traditional Acehnese buildings under the house, where the poles stand, the body of the house through carvings of house ornaments, and the top of the house, namely the roof of the house facing the sea. Meanwhile, fibers and pegs also have an important role. Each of the parts that make up traditional Acehnese houses is connected by continuous connections and reinforced with fibers and pegs without nails. This makes the house flexible (flexible and not rigid) that it can follow the direction of the earthquake movement. Each part supports the other to defend its construction against earthquake shocks.

Besides, even the traditional Acehnese houses do not use iron nails or any materials that can make the house heavy. The ends of each beam are joined with dowels and then reinforced with each other by chiseling and holes. In addition, the number of pillars supporting Rumoh Aceh is also large. So that these components can make the house stronger. Even to withstand earthquakes, the poles used are selected from solid wood. Not only that, the position of the pole is also arranged not too tightly. This is very useful to provide space for water currents to pass under the house more freely. So that this traditional Acehnese house is very safe and planned to anticipate flooding.

3.2. The existance of stilt house in Palu

Palu is the capital city of Central Sulawesi Province, Indonesia. Palu is a city located in Central Sulawesi, bordering Donggala Regency on the west and north, Sigi Regency in the south, and Parigi Moutong Regency in the east (Wikipedia, 2020). The people of Palu had already lived with earthquakes. Therefore the local culture has also reflected efforts to anticipate when a disaster occurs, including in designing buildings. The building model that is considered to be able to withstand a disaster is a stilts house. This stage model has various functions. One of them is to prevent wild animals from entering the house. For public buildings, the stage under the house can be used as a rest area. People also often use it as an animal enclosure.

The design of the stilts house in Palu has the advantage of being able to withstand floods and tsunami. However, the waves come are not as body highest of the building or above two meters from the ground floor. The house is also able to withstand a tsunami if the incoming water does not carry a large amount of debris. The house will be destroyed when large debris carried by the wave hits the legs of the building. After the tsunami hit Palu in 2018, most of the traditional houses of Central Sulawesi were still standing, even though the buildings on the left and right were damaged, and flattened to the ground.

One of the stilt houses that has survived and is still standing strong until now after being hit by the earthquake and tsunami in Palu is Banua Mbaso. This house was founded about 115 years ago, measuring 32 x 11.5 m, and was built by King Palu Jodjokodi around 1892, and is the residence of the King and his family and functioned as the center of the royal government at that time. This house on stilts is a combination of Bugis (South Sulawesi) and South Kalimantan-style architecture, which has 36 pillars of the main house and gandaria (terrace) including 8 kitchen poles.

Figure 7. Banua Mbaso

Figure 8. Banua Mbaso side view

The proportion of height between the three parts is as follows:
- The height of the foot is 2.4 m, lower than the middle at 2.88 m.
- The upper part is bigger in proportion, that is, the height from the attic to the top of the roof ridge is about 3.59 m.

In structural system of Banua Mbaso, there are:
1. The connection between the main floor beam and the main pillar of Banua Mbaso, the functions of the pole position not shifting.

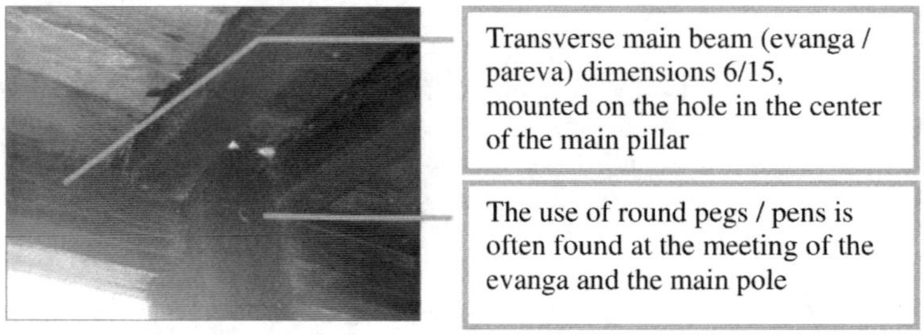

2. The connection between the *pelangga, evanga*, and the Banua Mbaso pole. These beams are floor beams that are fixed longitudinally or crossed with the main beam *(evanga / pareva)* making them stronger, the support beams are notched at the part where they meet the main beam.

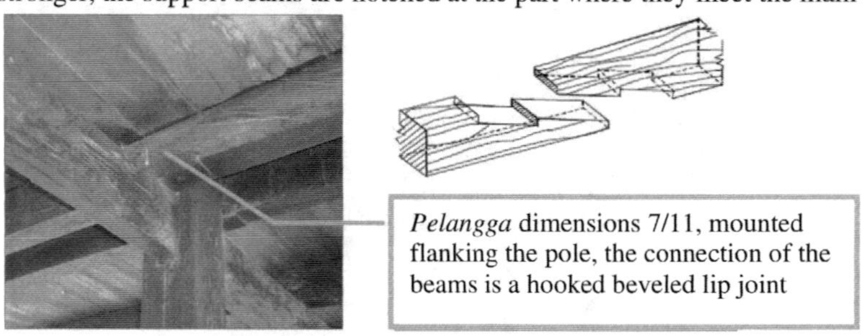

3. Sobo / tau-tau and floor. In the Banua Mbaso house, the floor is made of ironwood with a width of between 12-15 cm and a thickness of 3 cm, the length dimensions follow the pole intersection and support beams, except for Gandaria, Pakuntu, and Avu floors.

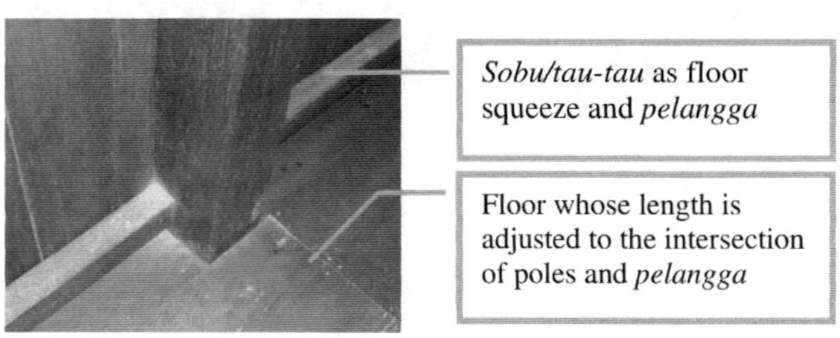

4. Connection of attic main beam and main pillar of Banua Mbaso. The roof beam is mounted in a transverse position, its position is adjusted to the hole in the vumbu. *Ulu avu* extends from the front of the Banua Mbaso roof to the back of the main house.

5. Roof. The joints in the truss generally use a grooved tongue with reinforcement with round pins or pins.

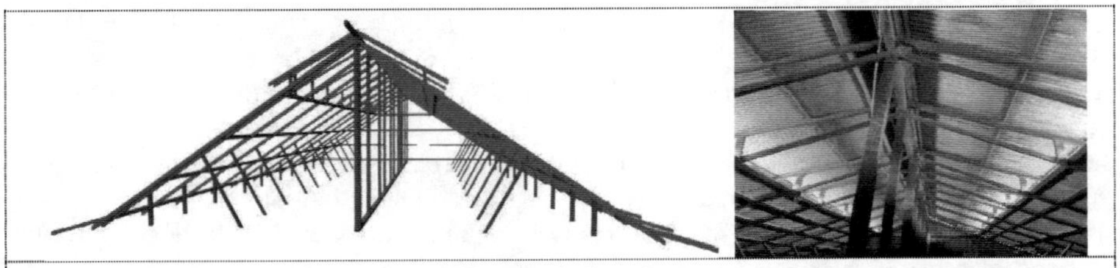

Figure 9. Roof truss

Apart from tsunami waves, traditional houses are also safe against earthquakes of a certain scale. The materials used are generally natural ingredients. Materials affect the structural strength of a building against shocks. In traditional buildings, generally using wood which has better flexural strength than concrete material. Wood is known for its flexural power, while concrete is very rigid. The material used is also selected or the highest quality wood. In addition, bonding the beams between the logs uses pins and ties, more flexible when hit by shocks. Almost all types of houses on stilts use stone as a reinforcing structure for the foundation. The stones are at certain points.

4. Conclusion

This research is not a new research anymore, but this research provides reinforcement to previous studies which stated that stilt houses are a type of house that can withstand natural disasters, such as earthquakes and even tsunamis. This is also evidenced by houses on stilts in Aceh and Palu after being hit by the earthquake and tsunami. The structure of the stilts house has been proven to be a type of building that withstands natural disasters such as earthquakes and even tsunami. The existence of under the floor creates a space to release vibration energy from the ground only a small part propagates to the physical house. Stilts house is built directly attached to the ground without a stage will receive energy directly from the earthquake so it is riskier than a stilts house. Earthquake vibration energy will directly propagate through the foundation of the building roof. Stilts house tradition has been owned by Indonesian people for a long time and is one form of how the community can withstand disasters. Houses made of wood, rattan, or bamboo are more flexible with vibrations thus minimizing damage and loss of life. This was seen in the two research locations in Aceh and Palu during the earthquake and tsunami, which made stilts house able to withstand tens of times when hit by an earthquake in a short period. The key lies in the strength of the building structure, the existence of a stilts house still exists even though it has been hit by a large-scale natural disaster.

5. References

[1] Daly, P., & Rahmayati, Y. (2012). Cultural heritage and community recovery in post- tsunami Aceh. From the Ground Up: Perspectives on Post-Tsunami and Post- Conflict Aceh. Singapore: Institute of Southeast Asian Studies, 57-78.

[2] Frick, I. H. (2004). Konstruksi Arsitektur 7 ILMU KONSTRUKSI BANGUNAN BAMBU, Pengantar Konstruksi Bambu (Vol. 7). Kanisius.

[3] Marwati, M. (2014). STUDI RUMAH PANGGUNG TAHAN GEMPA WOLOAN DI MINAHASA MANADO. TEKNOSAINS: MEDIA INFORMASI SAINS DAN TEKNOLOGI, 8(1), 95-108.

[4] Rofi, A., Doocy, S., & Robinson, C. (2006). Tsunami mortality and displacement in Aceh province, Indonesia. Disasters, 30(3), 340-350.

[5] St. Aisyah Rahman. (2012). Antisipasi gempa pada konstruksi rumah tradisional Banua Mbaso. Alauddin University Press.

[6] Syahabuddin, W., Hildayanti, A., & Kara, A. (2019, April). Gender Aspects Based on Anthropometric Norms at the Traditional House of Karampuang. In International Conference on Gender Research (pp. 599-605). Academic Conferences International Limited.

[7] Syamsidar. (1991). Arsitektur Tradisional Daerah Sulawesi Utara. Jakarta:Departemen Pendidikan dan Kebudayaan

[8] Tribunnews.com. 2018. Kekuatan Gempa Tsunami di Aceh 2004 dan Palu 2018: Dampak, Korban dan Dana Bantuan. Link https://manado.tribunnews.com/amp/2018/10/02/kekuatangempa-tsunami-di- aceh-2004-dan-palu-2018-dampak-korban-dandana-bantuan?page=all diunduh 9 September 2020

[9] Wasilah, A. H. (2018). Sistem Struktur Tiang Bengkok Pada Rumah Suku Kajang Bulukumba. Temu Ilmiah IPLBI 2018.

[10] Wasilah, W., & Hildayanti, A. (2017). Filosofi Penataan Ruang Spasial Vertikal Pada Rumah Tradisional Saoraja Lapinceng Kabupaten Barru. RUAS (Review of Urbanism and Architectural Studies), 14(2), 70-79.

[11] Widosari, W. (2010). Mempertahankan Kearifan Lokal Rumoh Aceh dalam Dinamika Kehidupan Masyarakat Pasca Gempa dan Tsunami. Local Wisdom: Jurnal Ilmiah Kajian Kearifan Lokal, 2(2), 27-36.

[12] Zulkarnain, A. S., & Hildayanti, A. (2018). INTEGRASI KONSEP ARSITEKTUR ISLAM PADA RUMAH ADAT SAORAJA LAPINCENG KABUPATEN BARRU. Nature: National Academic Journal of Architecture, 5(1), 1-12.

Local Traditions and Environmental Preservation - Genius Loci in *Penglipuran* Village in Protecting Public Spaces

I N U Mahardika,[1] Gusti Ayu Made Suartika,[1] and Kadek Edi Saputra[1]

1 Master Program in Architecture, Development & Planning of Udayana University – Bali, Indonesia

Email: udianamahardika10@gmail.com

Abstract. Community groups who live in traditional settlements mostly have a wealth called genius loci, which actually has very insight into the preservation of the natural and cultural environment. Bali also has several unique traditions that are not only in terms of settlement layouts, but are unique in terms of preserving the natural landscape. Meanwhile, there are some areas in Indonesia that are not able to maintain environmental sustainability, which is indicated by the continued increase in deforestation rates which are a sign of the deteriorating quality of the surrounding environment. Based on this, this article discusses the traditions applied in landscape conservation in Penglipuran Bangli village. This research was conducted using a qualitative descriptive method with a case study approach. To preserve the natural environment, there are several traditions that are still maintained and implemented consistently, including traditions: Preserving village forests, keeping the environment clean, telajakan karang kerti, may not use vehicles on the main village corridor. The application of local traditions related to environmental preservation is carried out consistently and with full awareness by both the community and the prajuru as leaders in the Penglipuran Traditional Village. The success of nature conservation cannot be separated from the clear, firm, consistent application of awig-awig accompanied by sanctions, as well as public awareness of the importance of landscape conservation.

Keywords: traditions, environmental preservation, public space, community participation

1. Introduction

Development in various sectors of life in order to improve people's welfare, on the other hand produces by-products in the form of waste and other waste materials. Development is a series of processes of optimization, independence and interaction between development components, namely natural resources, human resources, community values and technology [1]. Lessons can be drawn from local traditions in the form of values on how to utilize natural resources while taking into account the capabilities of nature itself. Local traditions also teach how waste products from resource utilization do not exceed the carrying capacity and carrying capacity of the environment it self.

Community groups who live in traditional settlements mostly have a wealth called genius loci, which actually has very insight into the preservation of the natural and cultural environment. But along with the times, local knowledge is getting eroded. This also applies when we talk about preserving the natural environment. For example, the Amazon Forest has been preserved for a long time, but then investors come, causing deforestation. Likewise with the Dayak forests in Kalimantan, the oil palm plantation industry is suspected to be the main cause of deforestation. During the period from 2011 to

2016, the amount of deforestation in Kalimantan was 2.08 hectares or 23 percent of the damage to national forests in that period [2]. The high rate of conversion of forest land every year has resulted in Indonesia being included in the list of countries that contribute to the eighth highest carbon emission in the world in 2018. Indonesia produces 965.3 million metric tons of carbon dioxide or equivalent to 2% of world emissions [3].

There are many things from local traditions that can be adopted and used as references for how urban planning should be structured to respond to the challenges of modernization which tend to exploit nature with high intensity. Some traditions which are local wisdom that develop in certain communities teach how to coexist with nature and the environment without destroying it [4,5]. For example, what is owned by the Panca Goya Tradition in Maluku, namely the tradition of cleaning hills, mountains and sacred places. Local people believe that these places have good fortune or are sacred so that they cannot be damaged. The Mantari Bondar tradition in North Sumatra, the surrounding community believes that the Batangtoru forest is a source of spring water, so its existence needs to be maintained and protected from generation to generation for the sustainability of generations. The traditional Sasi Nggama tradition in Papua is a unique tradition in preserving the environment and ecology, prohibiting the use of natural resources within a certain period of time where this tradition implies giving time to nature to be able to restore its natural resources [6].

Bali also has several traditional settlements that are unique, not only in terms of the layout of the settlements, but also unique in the form of local traditions in preserving the landscape environment. Among them are forest conservation efforts carried out by local communities in Tenganan Village, Karangasem Regency, Bali Province. With a strong tradition, the Tenganan community strives to maintain, maintain and prevent environmental damage through customary rules called awig-awig. Awig-awig in Tenganan Village, among others, are regulations related to the prohibition of logging several types of trees in the village forest as well as regulations related to the taking of wood materials that can be used by residents. The tradition of landscape protection carried out by the people of Tenganan is an effort to defend their ancestral heritage, especially from pressure from outside investors. In addition to Tenganan Village in Bali, there is Penglipuran Village which still maintains its culture and traditions from generation to generation which are considered to be native Balinese [7,8]. The uniqueness of Penglipuran Village lies in the social system, culture as well as the physical structure and spatial pattern of the village which is distinctive with a strong north-south linear pattern. Where the edge of the village is surrounded by village forests in the form of bamboo forests, wood forests and tegalan as a green open space so that it reflects the beauty of a truly Balinese village.

The Penglipuran Traditional Village community is one of the villages in Bali that persistently maintains their identity and continues their tradition in preserving their natural environment. While in some areas in Indonesia are not able to maintain environmental sustainability. For example, the continued increase in deforestation rates as a marker of the deteriorating quality of the surrounding environment. In 2000 forest cover in Kalimantan was approximately 33.2 million hectares, then decreased to 28.3 hectares in 2009 and finally 24.8 million hectares in 2017 [9]. Based on this, this article focuses on talking about how traditions are applied in landscape conservation in the village of Penglipuran Bangli. The number of spaces in Penglipuran village, both public and private spaces, are regulated in such a way by the traditions they have inherited.

Various previous studies related to environmental conservation have been carried out in Penglipuran Village. This study focuses more on how to apply local traditions in controlling and maintaining the sustainability of the natural environment in Penglipuran Village, Bangli. The increasingly complex relationship between development and environmental preservation requires alternative solutions as a form of environmental conservation pattern. In this context, Penglipuran Village was chosen as a case study.

IOP Conf. Series: Earth and Environmental Science **903** (2021) 012010 doi:10.1088/1755-1315/903/1/012010

2. Literature Review

2.1 Local Traditions, and Spatial Planning Controls

Local traditions are inherent habits that are believed and carried out for generations by the local community [10]. The Balinese Aga community has unique local traditions related to its culture, settlement structure and government system [11]. Customary law, better known as awig-awig is a basic guideline as a rule of customary village in its government [12]. The hierarchy of Balinese customary law consists of Awig-awig which functions as a kind of law while Perarem functions as implementing regulations [13]. In Bali itself, the customary norms, customs or traditions that apply in a certain community group that are inherited from generation to generation are called dresta [14]. For Penglipuran Village, in the preservation of its natural environment there is harmonization between customary law and local traditions in preserving the nature that was inherited by their ancestors dynamically along with the times.

The concept of residential space in Bali is strongly influenced by the beliefs and values of local life. The life values that apply to each Traditional Village affect the diversity of the concept of space. For Bali Aga villages, in general, the physical structure and spatial pattern of the village are typical with a north-south linear pattern known as the upstream-teben concept. Research on the concept of upstream-teben has been carried out by many experts [15-17] stating that hulu-teben is the concept of space as a symbol of respect for ancestors which is oriented up-down, the mountain represents the top and the sea represents the bottom. In this case, in Penglipuran Village, the spaces are clearly and firmly divided including hulu as a sacred space, a sacred space where the Kayangan Desa Temple is located, teben where setra (village graves) are located, between upstream-teben there are residents' settlements. Uniquely, the village area is surrounded by village forests in the form of bamboo forests and fields.

To maintain the existence of traditional spatial planning, controlling measures and instruments are needed [30, 31, 32, 33, 34, 35]. Control provisions in spatial planning are essentially aimed at avoiding the occurrence of spatial use conflicts. Through good and planned control, it is hoped that the utilization of resource allocation can be managed optimally. In the end, through control, it is hoped that there will be harmony between the plan and its utilization, so that environmental damage due to development can be prevented or at least can be minimized [18]. As an instrument of spatial regulation, spatial control policies should contain clear rewards and punishments and the consistency of their enforcement [19]. Rewards are in the form of incentives as one of the instruments in controlling the use of space as an effort so that the implementation of development is in line with the predetermined plan.

2.2 Potentials and Challenges in the Inclusion of Local Traditions to Control Spatial Development and to mitigate Environmental Damages

Traditional communities basically have local traditions that function to preserve the surrounding environment. However, along with the issue of wrong modernization, many local wisdom values have been eroded. The misunderstanding of some people who think that traditional means backward has an impact on the abandonment of some traditions that have been carried out for generations. Research on the factors that influence changes in traditional Balinese settlements has been carried out by many experts [20-22]. Based on these studies, it can be concluded that changes are caused by internal factors and external factors. Internal factors include the community's need for new functions due to population growth, economic needs and changes in the socio-cultural view of society towards modernity. While external factors include opportunities from the tourism sector which is believed to be able to support the community's economy. At the same time which is slowly changing the mindset of the people from traditional which is persistent in maintaining its tradition to modern.

2.3 How Traditions are Accommodated and mediated in cgoverning spatial development and natural environment.

In fact, local traditions in spatial planning and preservation of the natural environment have been recognized by native Indonesians. Local traditions in spatial planning have divided the spaces according

IOP Conf. Series: Earth and Environmental Science **903** (2021) 012010 doi:10.1088/1755-1315/903/1/012010

to their designation based on the prevailing local values. Local traditions in controlling urban spatial development and preserving the natural environment can be accommodated by integrating them into regional/city regulations. In addition, these traditions can be accommodated through real traditional community participation in spatial planning and nature conservation. Accommodating local traditions means acknowledging the existence of customary law communities as required by the constitution [23].

In Bali, it is expected that the spatial planning process ideally adopts all the important elements contained in local wisdom which is a form of the union of tradition, culture and religion as a fortress to protect Balinese culture and nature [24]. So that regional planning related to spatial planning and preservation of the natural environment needs to accommodate local values, so that the plans drawn up and determined are easy to implement.

2.4 Community Participation in Sustaining Natural Environment

A person's knowledge of the environment is the result of a thought process based on experience interacting with the environment, so that a person tends to participate in environmental conservation as part of the environment where he lives and where he does all his activities [25]. In connection with this article, the level of community participation in preserving the environment is inseparable from the availability of local values related to the environment and how these values are applied.

Sherry Arnstein (1969) in Suciati (2006), community participation is how people can be involved in social change that allows them to get a share of the benefits of influential groups [26]. There are three important elements in community participation in development to achieve optimal results, namely: the willingness of the community to be independent, the existence of mutually agreed goals and carried out with a planning, implementation and evaluation management system [27]. In this article, we will see how the participation of the Penglipuran community in carrying out the tradition of preserving their natural environment from these three aspects will be seen.

3. Method

This research was conducted using a qualitative descriptive method with a case study approach. The descriptive qualitative research method in this study was made with the aim of knowing how to apply local traditions in controlling and maintaining the sustainability of the natural environment in Penglipuran Village, in particular how local traditions are able to act as driving forces in maintaining environmental preservation. The case study approach is used to understand certain phenomena in a certain place and at a certain time. The data collection method used was by conducting a survey in Penglipuran Village and observing, namely interacting by looking at objects/subjects in the field, viewing, reading documents and interviews. While the data analysis method used to reveal the findings is qualitative data analysis.

4. Data and Disscusion

Penglipuran Village is one of the old villages in Bali or known as the Bali Aga. Bali Aga, also known as mountain Bali, are villages in Bali that are inhabited by indigenous Balinese people who are commonly called Bali Mula or Bali Kuna [28]. Penglipuran Village has a unique tradition, culture and physical structure of a clean and orderly village environment, so that it is able to display a sustainable rural face. The physical order of Penglipuran Village cannot be separated from the traditions and culture of the community that are rooted and have been passed down from generation to generation.

Penglipuran Village is located in Kubu Village, Bangli District, Bangli Regency, Bali Province, with an altitude of 500-600 meters above sea level and is located at GPS coordinates 8.0292893o Ls, 115.030360 East Longitude. Penglipuran Village is 5 Km from downtown Bangli and 45 Km from downtown Denpasar. Where in the east is directly bordered by the Kubu traditional village area, in the south by the Gunaksa traditional village, in the west by the Sang-Sang tukad while in the north it borders the Kayang traditional village (figure 1).

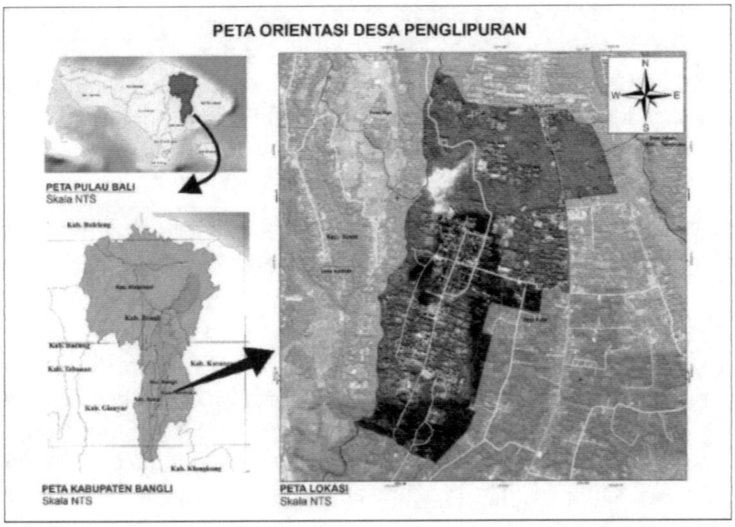

Figure 1. Map of Penglipuran Settlement [29]

Based on data from the Land Cover Map of the Penglipuran Priority Area in the Detailed Spatial Planning of the Bangli District, the area of Penglipuran Village is approximately 146.36 hectares. Utilization of village space, based on the land cover map of Penglipuran Village, is dominated by 48.45 hectares of village forest, 70.19 hectares of fields/moor, and 21.64 hectares of settlements. The environment has an important role in the life of the Penglipuran community. The tradition of preserving the natural environment can be seen from how the Penglipuran community persistently preserves the environment while maintaining spaces for various functions. In Penglipuran, village spatial planning consistently maintains its traditional spatial concept, namely hulu-teben. The natural environment that is preserved in Penglipuran village includes: village forest, dry fields, telajakan karang kerti in traditional settlements, setra (cemetery) and the war memorial (figure 2).

Figure 2. Penglipuran village spatial pattern map [29]

The village forest in the form of a bamboo forest is located upstream, precisely to the north of the village settlement. Telajakan is located on the main corridor of the village road which is located in the middle of the settlement and divides the settlement into two parts, namely on the west and east. Setra or graves are located in teben, located on the south end of the village.

4.1 Tradition in protecting natural environment in Penglipuran Village

To preserve the natural environment, there are several traditions that are still maintained and implemented consistently in Penglipuran Village, namely:

4.1.1 Village forest preservation tradition

Village forests in the form of bamboo forests belonging to traditional villages are managed by the community and the traditional villages themselves in the form of laba adat. Customary rules mandate that bamboo forests belonging to traditional villages managed by the community are not allowed to be converted. People who change the function of bamboo forests will be subject to customary sanctions according to the applicable awig-awig. The type and amount of the sanctions given depend on the results decided by the traditional village paruman (meeting). So far, no indigenous people have violated the regulation on the conversion of the village forest function.

Indigenous peoples are allowed to use bamboo plant products from village forests for their traditional house building needs and for commercial purposes by selective logging according to the forest lands they manage. There are 14 types of bamboo planted in the bamboo forest of Penglipuran Village, namely: *Jajang Aya, Jajang Bali, Jajang Panteg, Jajang Taluh, Jajang Papah, Jajang Batu, Tambang Gading, Tambang, Petung Buluh, Buluh, Tali Suet, Tali, Gading, Ampel* (figure 3).

Figure 3. Types and conditions of bamboo forests in Penglipuran Village (Authors' documentation, 2021)

Bamboo plants in Penglipuran Village have been maintained by the Penglipuran community for generations. This bamboo plant has a sacred value for the Penglipuran community because of its strategic location in the hulu (head) of the village. Apart from being a form of nature conservation, the preservation of bamboo forests for the Penglipuran community is a form of devotion and respect for the beliefs and teachings they hold dear. The felling of bamboo trees is also required to follow the customary rules that apply in Penglipuran Village. The felling of bamboo trees should not be carried out on the day of ingkelbuku and kajeng sweet. Local people believe that logging during ingkel Buku will cause the bamboo to break and break quickly, while logging when it is sweet is believed that bamboo plants will flower and bamboo plants will die quickly.

The regulation prohibiting the conversion of bamboo forest functions has become a standard rule that must be obeyed by all Penglipuran residents. This rule is based on, among other reasons, the function of the bamboo forest as a barrier to other villages, the bamboo forest as a catchment area and bamboo stalks as the main material for the Penglipuran traditional house. So that it can be seen clearly the relationship between conservation and the function of bamboo forest utilization in Penglipuran Village. The awig of preserving the bamboo forest was agreed to through the traditional village paruman. Awig is carried out and supervised by the community with the traditional warriors consistently. Communities who violate the rules prohibiting the conversion of bamboo forest functions will be subject to sanctions according to the results of the agreement in the traditional village paroman. The highest sanction for awig-awig violators in Penglipuran village is keneroyang (dismissed) as a Penglipuran traditional citizen.

The Penglipuran community maintains the bamboo forest other than as a form of conservation of nature, because bamboo trees are the main raw material for traditional buildings in Penglipuran village. These buildings are preserved because their functions and benefits greatly support the implementation of several traditional rituals in Penglipuran Village. Awig-awig in Penglipuran Village mandated that the Penglipuran Village community is obliged to preserve their traditional buildings in the form of: angkul-angkul buildings, umah paon buildings and bale dangin buildings (figure 4). Through this regulation, local people continue to preserve the bamboo forest in Penglipuran Village.

Figure 4. Traditional buildings of Penglipuran Settlement - Using Bamboo as the Main Building Material (Authors' documentation, 2021)

4.1.2 The tradition in keeping the environment clean

The tradition of keeping the environment clean in the Penglipuran village area is clearly and unequivocally contained in the awig-awig of the Penglipuran traditional village. In awig-awig, it is regulated that every traditional villager is not allowed to throw garbage indiscriminately that is not in accordance with the provisions of the traditional village. Violations of these rules in awig are subject to sanctions: "sang sapa sira mamurug pasukertan inucap kekanin pidanda manut perarem." Meaning: whoever violates these rules will be subject to a fine according to the perarem. This tradition is maintained because it has an impact on the beauty and sustainability of the village environment. Local people believe that apart from disturbing aesthetics and smelling bad, unmanaged waste can become a breeding ground for disease vectors and rodents. through good waste management, shows that there is an effort to maintain the quality of the environment, both air, water and soil in Penglipuran village.

The tradition of maintaining the cleanliness of the village environment is carried out by carrying out mutual cooperation every 15 days. Gotong-royong is done by cleaning the environment in Penglipuran village. The implementation of gotong royong is carried out in all village temples, neighborhood roads, cemeteries and all other public facilities. In village awig-awig, it is regulated that traditional villagers are obliged to maintain cleanliness at palemahan setre (cemetry area). The implementation of gotong royong is led by the Traditional Village Prajuru as the coordinator of customary manners in the administration of customary village governance. The tradition of keeping the environment clean is not only carried out in the implementation of mutual cooperation which has been determined periodically by the traditional village, but the community is aware and willing to clean up their surrounding environment. It is a tradition that has become a tradition in Penglipuran village where every day the community cleans the residential environment, starting from the yard of the house to the foothills.

IOP Conf. Series: Earth and Environmental Science **903** (2021) 012010 doi:10.1088/1755-1315/903/1/012010

To maintain environmental health, the traditional village awig-awig also forbids all Karang Kerti residents to dispose or channel the remaining liquid domestic wastewater into the drainage channel leading to the village's main road corridor. The rest of the waste water must be channeled into the drainage channel at the back of the Karang Kerti or absorbed at the back of the house yard or what is called premises tebe (open space at the back of the house). The awig-awig also includes the following: "*sapa sire ugi mekarya songbah mangde ngewentenang patemon bebaosan ring sang panyandingnya. Prade nenten adung patut kebaosin antuk prajuru adat utawi sang rumawos*". The awig-awig fragment means: whoever will build a sewerage channel to discuss it with neighbors, if no agreement is found, it will be discussed with traditional leaders. Local regulations in Penglipuran village mandate how the environment must be managed so that its sustainability can be maintained. The heaviest sanction for violators of this provision is the revocation of the right to use the Karang Kerti (house yard). Where Karang Kerti is managed by the traditional village.

4.1.3 Tradition in preserving of tejakan of the karang kerti

In addition to the habit of maintaining environmental cleanliness, villagers are also required to preserve the natural environment by maintaining the telajakan and natah space (open space in the yard between umah paon and bale dangin). The traditional village awig-awig explicitly mandates that there should be no transfer of the function of the yard yard for other functions. Uniquely, all the house yards in Penglipuran Village have beautiful treads with the same width and have been maintained until now (figure 5).

Figure 5. The condition of the bed and natah with various ornamental plants in Penglipuran Village (Authors' documentation, 2021)

In the awig-awig of the traditional village of Penglipuran, it is stated that " telajakan karang kerti sane majeng patut kapiara tur katandurin sarwa sari tan kengin katandurin tanduran siyosan sane ngawinang tan asri ring sajeroning margi." The customary rules mean that the garden of the house must be planted with flowers, not allowed to be planted with plants that cause disturbing the beauty along the road. In this case, all villagers who own coral reefs are obliged to participate in maintaining the cleanliness and beauty of their respective treads. Violations of these customary provisions will be subject to sanctions according to the regulations. The lightest sanction is the obligation to perform a pecaruan ceremony (traditional cleansing ceremony) at the village temple and the heaviest sanction is the revocation of the right to use Karang Kerti by the traditional village. Until now, this tradition is still carried out obediently by the residents, this is indicated by the absence of community members who have received customary sanctions for violating this rule.

4.1.4 Vehicles free public corridor

Customary provisions mandate that the main village corridor can only be passed by pedestrians. Vehicles in the form of motorized vehicles or bicycles are prohibited from crossing this main route. This tradition is seen as a form of preservation of the natural environment in Penglipuran Village. With this arrangement, you can feel the air around the main corridor of the village is very clean and fresh. This tradition is still maintained and preserved. The community's obedience in carrying out this tradition can be seen from the absence of violators of this rule. Penglipuran people who violate this tradition will be subject to sanctions in the form of fines according to the regulations. Perarem, in this case, is the result of an agreement that is discussed in the traditional village paruman (meeting).

4.2 Customary Codes and Regulations

Traditional village as a unit of customary law, bound by customs or customary law that grows and develops within the local community. Customary law, better known as awig-awig and perarem, is the basic guideline for traditional villages in their governance (Dharmayuda, 2001). In the context of Penglipuran Village, awig-awig and perarem have a major role in regulating the pattern of community trust relations (Parahyangan), social relations (Pawongan) and the sustainability of their natural environment (Palemahan). Penglipuran traditional village awig-awig which contains environmental conservation was agreed and ratified on August 19, 1989. Penglipuran village awig-awig regulates how environmental conservation is carried out and sanctions for violators of these rules.

The government system in Penglipuran Village consists of two government systems, namely the official government system and the traditional village. The government service system is under the leadership of the head of the environment who is tasked with coordinating affairs in the field of government administration. Meanwhile, the traditional village government is led by the Penglipuran Traditional Village Prajuru. The Penglipuran Traditional Village Government System consists of Prajuru Adat and Ulu Apad. Prajuru Adat is a democratically elected traditional village management structure, the management consists of the Kelian Desa adat, the Panglima, Pengarikan and the Interpreter Raksa/Petengen. Meanwhile, Ulu Apad is a traditional government system based on the order of seniority of cacakan krama Pengarep as part of the Bali Aga village. The structure of the Penglipuran traditional village institution can be seen in figure 6.

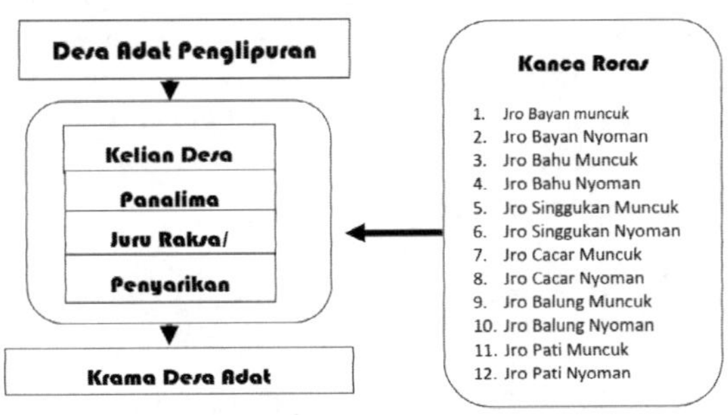

Figure 6. Leadership Structure of the Penglipuran Community (Authors' documentation, 2021)

In terms of environmental preservation, the application of awig-awig and perarem in Penglipuran Village is coordinated by the traditional Prajuru and the apparatus owned and supervised by Ulu Apad as an advisor to the institutional system of the Penglipuran Traditional Village. Indigenous peoples are obliged to comply with the agreed awig-awig. For people who violate customary provisions, especially those related to environmental conservation, they will receive customary sanctions according to the level and magnitude of the violation determined through the traditional village sangkepan (meeting). The customary rules related to environmental conservation are in line with the applicable state provisions.

4.3 How the existing rules are enforced, obeyed, violated, and sanctioned

In Penglipuran village, the tradition of preserving nature is in line with their customary law. In general, local regulations related to environmental conservation have been regulated in village awig-awig. But in its implementation, technically and in detail, it is regulated in the perarem. In Penglipuran Village, Perarem regulates in more detail how the implementation of awig-awig related to environmental conservation is enforced, adhered to and how to impose sanctions for violations of customary law. For

IOP Conf. Series: Earth and Environmental Science **903** (2021) 012010 doi:10.1088/1755-1315/903/1/012010

example, in Penglipuran Village's awig-awig, it is stated that customary village residents are obliged to maintain the palemahan (natural environment) of Penglipuran Village, manut dresta. this means that every villager is obliged to preserve the natural environment in accordance with the rules and traditions that apply in Penglipuran Village. Meanwhile, the Perarem regulates in detail the rights and obligations in the application of the awig-awig. For example, in the weakening conservation regulation, there is an obligation for indigenous people to protect the village forest in the form of a bamboo forest and it is prohibited to convert it to function. Indigenous people who have rights to use village forests are given management rights in accordance with applicable rules and traditions (manut dresta). The application of awig-awig and perarem in environmental conservation in Penglipuran Village is binding and forces indigenous people to obey them, because each of these rules is accompanied by sanctions for violators.

The rules in the form of customary law consisting of awig-awig and perarem are very effective in maintaining the sustainability of the traditional space and natural environment of Penglipuran village. Clear rules and strict implementation in their enforcement so as to make the people in Penglipuran Village very obedient to the mutually agreed rules. In formulating customary regulations, customary institutions involve indigenous peoples directly through the traditional adat meeting (customary meeting), so that all established customary rules can be accepted and implemented by the community with full awareness. In its implementation, the adat community is also a forum to evaluate the implementation of the preservation of the natural environment in Penglipuran Village.

In maintaining environmental preservation, Penglipuran Village applies clear and firm rules. Sanctions for violations of traditional village awig-awig are subject to criminal penalties, namely violations of customary rules are subject to sanctions based on perarem. The lowest customary sanction for indigenous peoples who violate customary law is the obligation to carry out a traditional cleansing ceremony (melaburin sepura) at Pura Desa. Meanwhile, the heaviest sanction is keneroyang, which is being dismissed as a customary citizen in the form of liberation of rights and obligations from the traditional village. The application of sanctions is carried out through the traditional village paroman mechanism. So far, no indigenous people have violated customary provisions related to the preservation of their natural environment. The obedience of indigenous peoples in carrying out the rules in environmental conservation is due to the synergy between tradition and applicable customary law. What is no less important in Penglipuran Village is that traditions and rules are enforced consistently and with full awareness by traditional institutions and indigenous peoples.

4.4 The Success of Tradition-Based Rules in controling Spatial Development and its Natural Environment

The success of the regulation in the form of awig desa arranged in the context of environmental preservation can be seen from the consistency of the people in Penglipuran Village in maintaining spaces that have a function to support the preservation of the village environment. Until now, Penglipuran Village is still consistent in preserving its natural environment as an ancestral heritage in the midst of modernization. Until now, no residents have received sanctions due to violations of awig-awig related to the preservation of the natural environment in Penglipuran Village because of the attitude of residents who obey the rules and are full of awareness.

Penglipuran Village's success in preserving its environment can be seen from the various appreciations received by the Traditional Village from various parties. In 1995, the Indonesian government awarded the Kalpataru award for efforts to protect bamboo forests in traditional village areas. In 2016 on Trip Advisor in the form of Traveler's Choice Destination Penglipuran ranked second after the Galapagos Islands in Ecuador. In 2017 it received the best award in the category of cultural preservation in the ISTA (Indonesia Sustainable Tourism Award) and in 2018 in the Magazine of Bombastic, Penglipuran Traditional Village was named the three cleanest villages in the world along with Giethoorn Village in the Netherlands and Maylynnong Village in India.

IOP Conf. Series: Earth and Environmental Science **903** (2021) 012010 doi:10.1088/1755-1315/903/1/012010

4.5 The Role of Community Members and Their Tradition-based Local Leadership

Penglipuran Village community as the main fortress in preserving the natural environment. The community is also the main actor of development in the village, especially in the field of environmental conservation, this can be seen from the high participation of the community in all stages of implementing activities related to environmental conservation in the village. The form of community participation in preserving the natural environment is carried out with the consistency of residents in maintaining environmental preservation traditions that exist in Penglipuran Village.

In the activities of preserving the natural environment, the Penglipuran village community plays an independent role with full awareness. The forms of participation given are in the form of energy, material, thought. The community is active in conservation activities, both on the agenda of the traditional village and in daily life. The community is fully involved in every conservation activity starting from the planning, implementation and evaluation stages. The role of the community in planning is carried out by providing ideas in the traditional village paruman (meeting). The community participates in determining the goals of conservation and how these goals are achieved as outlined in the customary village perarem. The implementation of environmental conservation is evaluated in the traditional village paruman activity which is carried out once every 15 days. The results of the evaluation in the traditional village paroman are used as the basis for making plans to improve the implementation of environmental conservation. The high level of community participation is inseparable from the participation of soldiers in traditional institutions. Prajuru coordinates the implementation of environmental conservation in the customary village area actively and consistently referring to the awig-awig of the traditional village. In Penglipuran Village, the prajuru are not only leaders but also those who are elders in adat, so that all environmental conservation activities coordinated by the prajuru will be followed by the villagers.

Penglipuran Village teaches how local traditions can act as a fortress to preserve the natural environment in Penglipuran Village. Local traditions, especially related to environmental preservation, are maintained and carried out consistently by all components of society. The local community is significantly involved in preserving the village environment, including in setting regulations in the form of awig-awig and traditional village perarem. Traditions relating to the preservation of the natural environment that have been successfully implemented must be adopted by integrating the values contained in these traditions into spatial planning documents. compiled by the government.

5. Conclusion

Local traditions in preserving the natural environment in Penglipuran Village are able to preserve the environment. This can be seen from the preservation of spaces that have a protective function in Penglipuran Village. Spaces that have protected functions that are still preserved include: village forest in the form of bamboo forest, village moor, telajakan Karang Kerti in traditional settlements, setra and hero's grave garden. In nature conservation, these spaces function as catchment areas and green open spaces. The preservation of these spaces is inseparable from the existing landscape preservation traditions, including: Preserving village forests, maintaining environmental cleanliness, preserving coral reefs, not using vehicles on the main village corridor. In its implementation, the application of tradition is carried out with full awareness by all components of society, both by indigenous people and prajuru as leaders in the Penglipuran Traditional Village. The success of nature conservation cannot be separated from the clear, firm, consistent application of awig-awig accompanied by sanctions, as well as public awareness of the importance of landscape conservation. The application of customary law in conservation is very important, but far more important than that is the awareness of the Penglipuran traditional village community about the importance of landscape preservation to support the lives of today's people and for generations to come.

References

[1] Muta'ali 2012 *Daya Dukung Lingkungan Untuk Perencanaan Pengembangan Pengembangan Wilayah* (Yogyakarta: Badan Penerbit Fakultas Geografi UGM)

[2] Kemen G A et al 2019 *Environ. Res. Lett.* 14 024007
[3] Pusparisa Y Databoks [internet] *Sepuluh Negara Penyumbang Emisi Gas Rumah Kaca Terbesar* [update 2021 Februari 16 cited 2021 August 24] available at: https://databoks.katadata.co.id/datapublish/2021/02/16/10-negara-penyumbang-emisi-gas-rumah-kaca-terbesar
[4] Sufia R et al 2016 *J. Pendidikan: Teori, Penelitian dan Pengembangan* kearifan lokal dalam melestarikan lingkungan hidup 1 4
[5] Swandi I W 2017 *J. Kajian Bali* kearifan lokal bali untuk pelestarian alam: kajian wacana kartun - kartun majalah "bog-bog" 07 02
[6] Wijanarko Tempo [internet] *Mengenal Sasi Nggama Tradisi yang menjaga lingkungan Kaimana.* [cited 2021 August 02] available at: https://travel.tempo.co/read/1033521/mengenal-sasi-nggama-tradisi-yang-menjaga-lingkungan-kaimana
[7] Reuter T A 2005 *Custodias of Sacred Mountains: Budaya dan Masyarakat di Pegunungan Bali* (Jakarta: Yayasan Obor Rakyat)
[8] G A M Suartika 2018 *IOP Conf. Ser.: Earth Environ. Sci.* 123 012014.
[9] Ramayanti F CNN Indonesia [internet] *Hutan Kalimantan Hilang, Banjir Menerjang* [cited 2021 August 28] availabel at: https://www.cnnindonesia.com/nasional/20210118140024-20-595141/hutan-kalimantan-hilang-banjir-menerjang
[10] Seitel 2001 *Intangible Cultural Heritage-working definitions* (Piedmont, Italy: UNESCO)
[11] Suada I N 2003 *Beberapa Desa Unik di Bali* (Bali: Dinas Kebudayaan Provinsi Bali)
[12] Dharmayuda I M 2001 *Kesatuan Masyarakat Hukum Adat di Provinsi Bali* (Denpasar: Penerbit Upada Sastra)
[13] Windia W. Kedudukan awig-awig dalam masyarakat hukum adat. Presented at Pelatihan Penyusunan Awig-Awig National Education University 2017 June 12; Bali Indonesia.
[14] Bali Province Government 2019 *Peraturan Daerah Provinsi Bali Nomor 4 Tahun 2019: Desa Adat di Bali.* (Bali: Lembaran Daerah Provinsi Bali).
[15] Ardana I G N G et al 1982 *Sejarah Perkembangan Hinduisme di Bali.* (Denpasar: Proyek Inventarisasi dan Dokumentasi Kebudayaan Bali, Departemen Pendidikan dan Kebudayaan Propinsi Dati I Bali)
[16] Gelebet et al 1982 *Arsitektur Tradisional Daerah Bali.* (Denpasar: Proyek Inventarisasi dan Dokumentasi Kebudayaan Bali, Depdikbud DT I Bali, Denpasar.
[17] Parwata I W and Nurwarsih 2015 *Konsep Topografi Alami "Hulu- Teben" di Banjar Gunungsari Desa Kawasan Wisata Desa Jatiluwih, Tabanan,* Prosiding Seminar Nasional Tata Ruang dan Space 2: Memastikan Penataan Ruang Yang Bekelanjutan: Kearifan Lokal dan Budaya Dunia dalam Penataan Ruang. (Denpasar: Universitas Hindu Indonesia).
[18] Muta'ali L 2013 *Penataan Ruang Wilayah dan Kota: Tinjauan Normatif-Teknis.* (Yogyakarta: Badan Penerbit Fakultas Geografi-UGM)
[19] Budiharjo E 2014 *Reformasi Perkotaan, Mencegah Wilayah Urban Menjadi Human Zoo.* (Jakarta: Kompas).
[20] Ganesa et al 2012. *Arsitektur e-Journal.* Pola ruang permukiman dan rumah tradisional bali aga Banjar Dauh Pura Tigawasa **5** 2.
[21] Mendra W and Wiriantari F 2016 *Jurnal Anala* Perubahan spasial permukiman tradisional di Desa Adat Tenganan Pegringsingan Bali **4(2)** 73-97.
[22] Agusintadewi K and Manik W Y 2016 *Menguak Tradisi Masyarakat Desa Adat Bali Aga di Kabupaten Bangli* (Bali: Udayana University Research Report).
[23] Kristyanto N 2017 *Jurnal RECHTSVINDING* kedudukan kearifan lokal dan peranan masyarakat dalam penataan ruang di daerah.**6(2)** 151-169.
[24] Muhaimin 2018 *Jurnal Penelitian Hukum DE JURE* Kedudukan kearifan lokal dalam penataan ruang Provinsi Bali. **18(1)** 59-72.
[25] Ariwidodo E 2014 *Nuansa* Relevansi pengetahuan masyarakat tentang lingkungan dan etika lingkungan dengan partisipasinya dalam pelestarian lingkungan. **11(1)** 1-20.
[26] Arnstein, Sherry R 1969. *Journal of the American Institute of Planners* A ladder of citizen participation. **35(4)** 216-224.

[27] Soetrisno L 1995 *Menuju Masyarakat Partisipatif* (Yogyakarta: Kanisius).

[28] Cuthbert A R and Suartika G A M 2017 *Jurnal Kajian Bali* Revisiting Reuter: symbolic and material economies in Bali aga society **7(1)** 259-298.

[29] Dinas Pekerjaan Umum Penataan Ruang Perumahan dan Permukiman 2019 *Rencana Detail Tata Ruang (RDTR) dan Peraturan Zonasi (PZ)* Kecamatan Bangli (Bangli: Dinas Pekerjaan Umum Penataan Ruang Perumahan dan Permukiman).

[30] Suartika, G A M (2013) 'Open Space Traditions, Development and Modernity: The Case of Bali' *Journal Nasional Terakreditasi Mudra: Journal Seni Budaya* Vol. 28, No. 2 December 2013, Institut, Seni Indonesia Denpasar, UPT Penerbit, p: 283-299.

[31] Suartika, G A M (2010) 'Substansi Budaya dalam Kebijakan Tata Ruang di Bali' *Journal Nasional Terakreditasi Humaniora: Jurnal Budaya, Sastra, dan Bahasa* Vol. 22, No. 3, Oktober 2010, Fakultas Ilmu Budaya Universitas Gadjah Mada, p: 313-326.

[32] Suartika G A M. (2009) Tenganan, Bali Aga and Development: A Socio-spatial Analysis of Culture,' *Journal of Architecture & Planning Studies*, Vol. 3 No.2, October 2009, p: 44-58.

[33] Suartika, G A M. Cuthbert, A R, Putra, G M, Saputra, K E (2020) 'Public Domain and Cultural Legacy: The Governance of a Sacred and Vernacular Cultural Landscape in Bali' *ISVS e-Journal* Vol 7, Issue 2, p:1-22.

[34] Suartika, G A M, Cuthbert, A R, (2020) ' Sleight of Hand: The Expropriation of Balinese Culture ' *Journal Space and Culture,* Online edition, Sage Publisher, p:1-21.

[35] Suartika, G A M, Zerby, J, Cuthbert, A R, (2018) 'Doors of Perception to Space – Time - Meaning: Ideology, Religion, and Aesthetics in Balinese Development' *Journal Space and Culture,* Vol 21 (4) November, p: 340-357, Sage Publisher.

Acknowledgement

Authors are grateful to the Indonesian Ministerial for Education, Culture, Research and Technology for enabling the publication of this article by providing research grants (with a contract Number of: B/136-10/UN14.4.A/PT.01.05/2021, dated 29 June 2021).

Conversion of Public Green Open Spaces to Facilitate the Tourist Industry

I Gusti Ayu Dwi Muliasari,[1] Gusti Ayu Made Suartika,[1] Kadek Edi Saputra[1]

[1] Masters Program in Architecture, Development and Planning, Udayana University-Bali, Indonesia.

Email – ayudwiwduya@gmail.com

Abstract. Green open space is a space is dominated by an open environment, it can be in the form of parks, green lines, or recreational facilities. The benefits of having green open space in an area are that it can improve the quality of the environment, create environmental harmony and balance, provide freshness, a place for water absorption, and as a component of the structure and space of the city. In the development of an area, the ecological value of the land is often sacrificed to be replaced with a more economical function, such as the construction of additional facilities in the tourism area. To maintain the benefits of green open space, this study discusses the increasing use of green open space in the construction of tourism supporting facilities, especially those in Giriloyo Batik Village. This research is expected to be used as a consideration for the management of the Giriloyo Batik Village area, which later on the area can be managed properly, both in terms of physical and non-physical. This study used qualitative research methods. The impact of the construction of tourism support facilities in Giriloyo Batik Village shows that in the last 14 years there has been a change in land use. In particular, the change in the use of green open space as a support for tourism areas of 3 hectares or 3% of green open space is now used as built-up land. The tourism facilities in question include Gazebo Batik, cottage industries, dam, parking areas, and secretariats that put pressure on the land.

Keywords: Public green open space; tourist development, land use conversion

1. Introduction

Green open space is an elongated area or pathway and grouped, whose use is more open, where plants grow, both those that grow naturally and those that are intentionally planted [1]. Green open space is generally intended for reforestation as one of the elements of the city which is determined by the comfort and beauty factor for a city space. The comfort in question can be in the form of silencers, shade or protective equipment from the hot sun, and neutralizing the air. Meanwhile, beauty can be in the form of plant arrangement assisted by constructions aimed at resisting erosion. The arrangement of green open spaces also applies the principles of beauty and comfort [2].

According to the 2006 Directorate General of Spatial Planning, green open space is a field overgrown with various plants, in various strata, ranging from ground cover, shrubs, shrubs, and trees. Balancing the natural environment with the artificial environment, namely as a "vendor" of the function of environmental sustainability in water, soil and air media as well as the conservation of flora and fauna

IOP Conf. Series: Earth and Environmental Science **903** (2021) 012011 doi:10.1088/1755-1315/903/1/012011

biological resources. The continuity of natural ecological functions in the urban environment in a balanced and sustainable manner will form a healthy and humane city [3].

Planning for green open space or green open space in an area must take into account the spatial capacity considerations in responding to pollutant, carrying-capacity, and land-capability [4], [5], [6], [7], [8]. The next consideration is human activities in utilizing space with increasing space requirements by developing community, capital, natural resources, and the environment [3]. All of these considerations are important to realize to achieve balance, harmony, and safety of open space and the surrounding environment. In addition, green open space planning in an area also needs to consider the realization of a harmonious, balanced, and in harmony with the surrounding environment. As part of the spatial plan, green open space in urban areas also has ecological, social, economic, and architectural functions and has aesthetic values that can make the value of pride and identity of the city itself [9].

Initially, most of the land in urban areas was open land or green open space. However, along with the increasing need for space in urban areas, these spaces are converted into built-up land. Development in an area tends to be accompanied by the development of the area. The development and development of an area is a process of land conversion to meet the needs of the community for the infrastructure and facilities needed in the development of an area [10].

In the development of an area, the ecological value of the land is often sacrificed to be replaced with a more economical function, such as the construction of additional facilities in the tourism area or land cover to support the accommodation of the area. The decrease in the quantity and quality of open space in urban areas has resulted in a decrease in environmental quality such as the occurrence of floods, increased air pollution, decreased community productivity due to stress due to the limitations of public spaces that are commonly used to interact [11].

The development of supporting space for the tourism industrial area can lead to a decrease in environmental quality, where with this development there will be a conversion of land functions that were originally green land into built-up land [12]. This land use change generally occurs in economically developing areas, such as what happened in the Giriloyo Batik Village. The changes that occur are caused by the development of tourism supporting infrastructure, the construction of tourism accommodation and its supporting facilities, and the conversion of land functions due to the existence of new settlements.

One of the related studies that has been carried out by Latifa Muliawati [13] in her thesis entitled Karangkulon Hamlet Tourism Landscape Planning in the Batik Village of Wukirsari Imogiri, Yogyakarta, discusses the character of the Karangkulon Hamlet landscape, both in terms of physical and socio-cultural aspects. The subject of this paper is the regional landscape planning which consists of spatial planning, circulation, vegetation, activities, and tourist facilities. The plan was developed based on the existing concept, namely the development of the Karangkulon Hamlet landscape into a batik tourism area that provides experience and education for visitors by considering the physical condition of the area that is prone to landslides. Where the planning is expected to be one of the efforts in improving the welfare of the community through tourism activities and preserving the batik culture in Karangkulon Hamlet [13].

Furthermore, research conducted by Ratika Tulus Wahyuhana and Mohammad Muktiali [14] with the title of research namely The Effect of the Sendang Asri Tourism Area of Gajah Mungkur Reservoir on Land Use Changes and Socio-Economic Aspects of the Community. This study uses quantitative research methods with sampling techniques (random sampling and purposive sampling). From the research that has been carried out by researchers, it can be concluded that the development of the Sendang Asri tourist area of Gajah Mungkur Reservoir has an influence on land use change which is marked by the development of tourist attractions such as water parks and entertainment stages as well as increasing trade and service areas. In the socio-economic aspect of the community, it is marked by increased knowledge of local communities in promoting local culture that is able to change income levels and provide additional income for the community in the tourist area [14].

Subsequent research was written by I Wayan Tagel Sidarta (2002) in his thesis entitled The Impact of Tourism Development on Environmental, Social, and Economic Conditions of the Community (Case Study of the Sanur Tourism Area, Denpasar, Bali). This study uses a sampling method with descriptive analysis. The purpose of this study is to (1) analyze the impact of tourism development in the Sanur tourism area on changes in employment and income, patterns of division of labor, employment opportunities, and doing business in the Sanur coastal area, (2) analyze the impact of tourism development on land use changes that occur. implications for the environment, and (3) formulating a planning approach to the management of the Sanur tourism area. The conclusion of this study is that there are three impacts of tourism development in the Sanur area including changes in employment and income, there is a pattern of division of labor to wives, and increased opportunities to work. In addition, the impact of the development of tourism is that there is a change in the function of land which was originally rice fields, gardens, and settlements turned into hotels, restaurants, and tourism supporting facilities [15].

So, in the context of this study, we will discuss the increasing use of green open space in the construction of tourism supporting facilities with Giriloyo Batik Village as a case study. Giriloyo Batik Village is one of the tourist destinations in Karangkulon Hamlet, this village is growing very rapidly. However, when it develops, the occurrence is inversely proportional to the availability of green open space. The rapid growth of tourism in Karangkulon Hamlet has converted land that was previously used as green open space, and this condition is very worrying if it occurs on an ongoing basis. By looking at this phenomenon, it is very important to write about increasing the use of green open space in the construction of tourism supporting facilities, where this study has never been done in Giriloyo Batik Village.

2. Theory Review

This study uses some of the literature on green open spaces. Starting from green open space as a supporter of balance in urban areas to the negative impacts of green open space conversion and strategies to overcome them. These literatures are taken from books as well as several previous studies.

According to Purnomohadi [3], green open space is a stretch of open land without buildings that have certain sizes, shapes and geographical boundaries with any status of control. Inside the green open space there are perennial woody plants, with trees as the main characterizing plants. In addition, there are other complementary plants such as shrubs, shrubs, grasses, and other ground cover plants. These plants are used as a complement and support for the function of the green open space in question [3].

According to Roger Trancik [16], green open space is a space or land that is dominated by an open environment, which can be in the form of parks, green lines, or recreational facilities [16]. According to Arianti [17], physically green open space can be divided into two, namely natural and non-natural or built. Where natural green open space can be in the form of wild habitats, protected forests, and national parks, while non-natural green open spaces can be in the form of parks, green roads, and also cemeteries. When viewed from the side of its function, green open space can function ecologically, socially, culturally, economically, and architecturally [16]. The function that is directly related to this research is the socio-economic function. For green open space with socio-economic functions, such as sports facilities, playgrounds, or rest areas, it must have good accessibility, including accessibility for persons with disabilities [17].

The benefits of having green open space in an area are [10], [18]:

1. Improving the quality of the living environment so that an environment can grow into a comfortable, fresh, beautiful and clean environment. Where if an environment has poor groundwater quality and has a high level of air pollution, it will cause various environmental impacts that are detrimental to society
2. Creating harmony and balance in the ecosystem environment that is useful for the benefit of the community through the realization of a city physical environment that prioritizes green open spaces

3. Providing freshness through the abundance of O2-producing plants that can maintain air circulation in an area as a means to improve the micro-environmental climate, provide comfort for the community because with lots of green open space it can reduce stress levels in urban communities and provide beauty in an area.
4. As a container for water absorption so as to reduce the risk of flooding
5. As a component forming the structure and spatial planning of the city which is very much needed in controlling the development and development of a region

The development of the tourism industry in developing countries that has not been regulated in such a way in an area tends to get rid of existing productive land, for example what happened in Yogyakarta. Based on data from the Central Statistics Agency for the Special Region of Yogyakarta in 2016 (Table 1) for the last ten years there has been a shift in the function of rice fields. In the period 2006 to 2015 there has been a conversion of rice fields of 3,089 hectares with an average converted area of 308.9 hectares per year. This figure is a very large number when compared to the number of rice fields clearing that occurred in the same period, which was 424 hectares for 10 years or 42.4 hectares per year. The conversion of paddy fields which is greater than the clearing of paddy fields indicates the potential for large loss of food production, so this can pose a threat to the food security of the population [19].

Table 1. Area and Rate of Transfer of Rice Fields in the Special Region of Yogyakarta (2006-2015)

Year	Rice Field Area (Ha)	Area of rice field land conversion (Ha)	Rice Field Clearing Area (Ha)	Depreciation Rate of Rice Field Area (%)
2006	56.218,00	0,00	0,00	0,00
2007	55.540,00	678,00	0,00	-1,21
2008	55.332,00	208,00	0,00	-0,37
2009	55.325,00	7,00	0,00	-0,01
2010	55.523,00	0,00	198,00	0,36
2011	55.291,00	232,00	0,00	-0,42
2012	55.110,00	181,00	0,00	-0,33
2013	55.336,00	0,00	226,00	0,41
2014	54.417,00	919,00	0,00	-1,66
2015	53.553,00	864,00	0,00	-1,59
Total	**551.645,00**	**3.089,00**	**424,00**	**-4,84**
Average	**55.164,50**	**308,90**	**42,40**	**-0,48**

Source: [20]

In many cases these productive lands are part of green open space so that the conversion of productive land also has implications for the availability of urban green open spaces. Factors that affect the availability of green open space are related to the presence or absence of detailed allocations as outlined in the planning, the availability of incentives and disincentives, the formation of facilitators, and the existence of various innovative programs for reforestation [21]. By decreasing the existing productive land, it can reduce the availability of green open space. In addition, it can also reduce the level of economic income of community members in the area. Especially in the context of this research, batik craftsmen usually use teak leaves as natural dyes. In the context of tourism development and the conversion of green open space functions in this area, it is very clear. It can be seen in Giriloyo Batik Village, the batik craftsmen in this village use teak leaves as natural dyes. While this teak tree is found

IOP Conf. Series: Earth and Environmental Science **903** (2021) 012011 doi:10.1088/1755-1315/903/1/012011

in the green open space or productive land. If the open space in the area is lost, the teak trees will also disappear, which causes the batik industry to not run as expected. This condition is exacerbated by the existence of various investors who intimidate the community to sell their land in order to provide space for the construction of tourism supporting facilities. An example is in the Batik Gazebo area, where an area of 0.3 Ha which was previously a green open space has now become a built-up area that is used as a tourism support facility. Gazebo Batik is the center of tourism activities in Giriloyo Batik Village.
The negative impact of the increasing use of green open space in the development of an area, namely [22]:

1. Decreased production and productivity of agricultural land, where some forms of land conversion in an area are related to the existence of agricultural land. This event will result in the loss of agricultural land, symptoms of commercialization (buying and selling activities), and agricultural intensification. In the context of this research, this occurs in the decreasing number of teak trees which are commonly used as raw materials for natural batik dyes.
2. There is a gap in the demand for land and the availability of undeveloped land. Where the availability of land is increasingly decreasing, resulting in an increase in land prices. In the context of this research, this happened on the lands around the Batik Gazebo.
3. Declining environmental quality, such as noisy environment, high air pollution, natural disasters (floods/landslides), and low environmental quality. In the context of this study, this occurs in the state of clean water in the area. With the narrowness of the available land, it is increasingly difficult for people to find clean water. In addition, the activity of disposing of batik waste in the environment also makes the quality of the existing environment decrease [23].

The existence of green open space in an area can function properly if the use of space is adjusted to the existing plan. Such as saving protected areas to ensure the preservation of regional functions and reducing the risk of natural disasters. Thus, the existence of green open space is not only a complementary element in the planning of an area, but also as an element forming the spatial structure of an area [3]. Where the elements forming the spatial structure in an area include detailed aspects (buildings, circulation systems, open spaces, and urban infrastructure), aspects of urban layout (spatial patterns, environmental composition built on patterns around the area), and regulatory aspects (total planning). and city plans that show the dynamics of the urban area) [24].

3. Method

This research uses a qualitative approach with case studies. The use of a case study approach in this study aims to provide an in-depth picture of the increasing use of green open space as a tourism facility. This research cycle begins with selecting the object of research, then continues with interviews and direct observations at the location. Then collect the data obtained from the interview and observation sessions, then analyze the data [25].

4. Data, discussion, and results/findings

4.1. Research Site Overview

Giriloyo Batik Village is a center for hand-drawn batik in Southeast Yogyakarta, namely in Karangkulon Hamlet, Wukirsari Village, Imogiri District, Bantul Regency, Special Region of Yogyakarta (7°54'58.0"S, 110°23'54.5"E, Figure 1). Where the Giriloyo Batik Village itself consists of three hamlets, namely Cengkehan Hamlet, Giriloyo Hamlet, and Karangkulon Hamlet. Giriloyo Batik Village is not far from the Bantul Regency Government Center which can be accessed via Jl. National III, Jl. Argorejo, Jl. Sedayu-Gesikan, Jl. Sultan Agung, Jl. Bakulan Imogiri, Jl. Scouts, and continued to Jl Giriloyo.

Figure 1. Location Map of Giriloyo Batik Village.
Source: Google maps accessed on 7 September 2021

Karangkulon Hamlet is the center of the growth of tourism activity in Giriloyo Batik Village. Where in the hamlet there are several tourism supporting facilities such as a Batik Gazebo, showroom (marketing location for batik craftsmen's work), a secretariat, and a parking area as well as a home industry for hand-drawn and stamped batik. In addition, there are also several restaurants, grocery stores, and also gurah treatment, which is a traditional treatment using a mixture of srigunggu liquid, which is dripped through the nose with the aim of removing mucus. [26].

Based on the location review and interviews with the two informants, namely Ms. Khibtiyah (Coordinator for Services and Education of the Giriloyo Batik Village Community) and Mr. Basuki Tapip (Head of TU and General Affairs of Wukirsari Village), the following results were obtained:

4.2. The Origins of the Formation of a Tourism Area in Giriloyo Batik Village

The Giriloyo Batik Village is located near the Imogiri Tomb Complex which has an impact on the spread of culture. Where in the past, most of the residents of Karangkulon Hamlet became courtiers of the Yogyakarta Palace who were in charge of caring for the tombs of the kings of Yogyakarta-Solo. From there, interactions began to occur between the Keraton and the people of Karangkulon and its surroundings, then several figures from the Keraton's relatives gave jobs to the surrounding community, especially women as batik nyanthing workers. Initially this was the case, for centuries the residents of Karangkulon who practiced batik were still laborers and sold semi-finished batik to batik masters in the city center around the Yogyakarta Palace for generations [27].

After the great earthquake devastated Yogyakarta in 2006, together with the spirit of togetherness to rise from trauma and adversity, the community formed batik groups and received assistance from the government or NGOs. Where this assistance provides training and brings in several experts to improve skills in making batik ready and ready to sell and the ability to market it, it is no longer completely dependent on big bosses in the city. After that, Karangkulon Hamlet, Cengkehan Hamlet, and Giriloyo Hamlet became one to form a community called Batik Giriloyo and in 2007 won a muri record as the "Longest Written Batik Shawl" involving nearly 1000 batik craftsmen. After that, the Giriloyo Batik Village was officially named the Center for the Batik Handwriting Industry of the Yogyakarta Palace [27].

After the formation of the Giriloyo Batik Village, the area is increasingly crowded with tourists, especially tourists who want to learn batik and shop. This condition continues, causing the area to become crowded. Tourists from various regions continue to visit the tourism area, which causes many wishes to add facilities to support these tourism activities.

4.3. Land Use in Karangkulon Hamlet in 2006

Based on the monograph data of Karangkulon Hamlet in 2006, the use of residential land is 35 hectares or 35% of the total area of Karangkulon Hamlet. Utilization of space for rice fields and forests based on existing land use conditions in Karangkulon Hamlet in 2006 reached an area of 63 Ha or 63% of the total area of Karangkulon Hamlet. The agricultural land in the hamlet is in the form of rice in the rainy season, secondary crops and sugar cane during the dry season. In addition, it is not uncommon to find lands planted with teak trees, where the young leaves of the teak tree itself can be used as a natural dye, which is brownish red. Details of land use in Karangkulon Hamlet until 2005 can be seen in table 2 and figure 2.

Table 2. Land Use in Karangkulon Hamlet in 2006

No.	Aplication	Area of Aplication (Ha)	Precentage (%)
Fill Area			
1.	Settlement	35	35
Total		**35**	**35**
Open Green Space Area			
1.	Riceflied and Forest	63	63
2.	Grave Yard	2	2
Total		**65**	**65**
	Total Land Use	**100**	**100**

Source: [28]

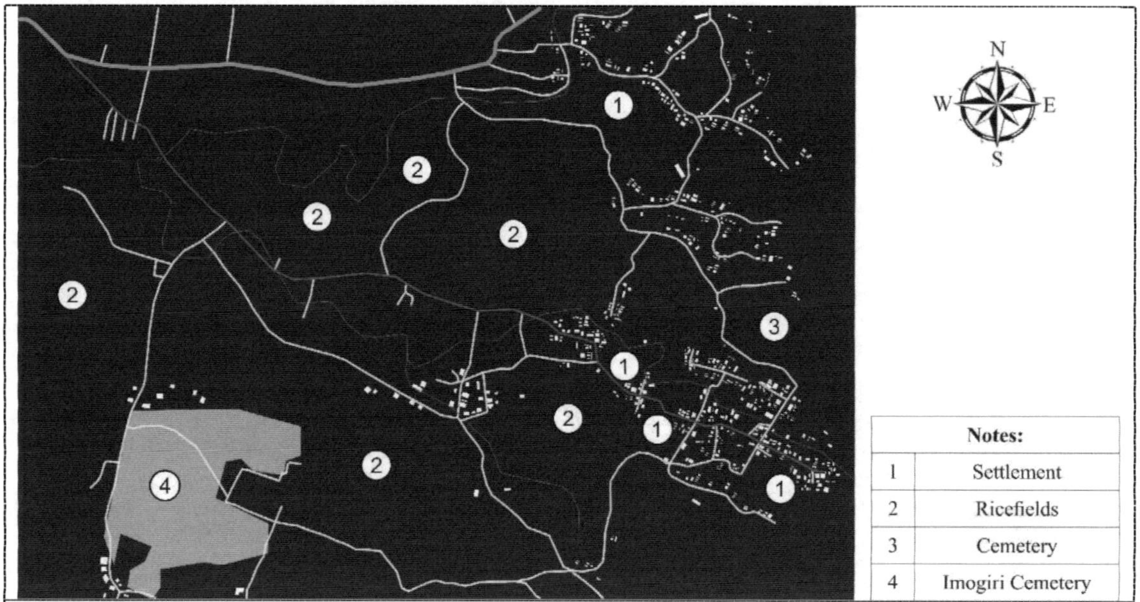

Figure 2. Mapping of Land Use at Karangkulon Hamlet 2006
Source: Google maps and data obtained from two informants

4.4. Land Use in Karangkulon Hamlet in 2021

Based on the monograph data of Karangkulon Hamlet in 2021, the condition of the existing land area is still the same, namely 100 Ha. However, there has been a development or change in the function of the land which was originally in the form of green open space (rice fields and forests) which has now turned into a variety of building functions. Where as much as 3 Ha or 3% of green open space is now used as built-up land including the Batik Gazebo (5), home industry (6), parking and secretariat area (7), and Embung Imogiri (8) (Figure 3).

Table 3. Land Use in Karangkulon Hamlet in 2020

No.	Aplication	Area of Aplication (Ha)	Precentage (%)
Fill Area			
1.	Settlement	35	35
Total		**35**	**35**
Open Green Space Area			
1.	Riceflied and Forest	60	60
2.	Grave Yard	2	2
Total		**62**	**62**
Land Use Change (Green Open Space Built-up Land)			
1.	Dam	2.5	2.5
2.	Gazebo Batik	0.3	0.3
3.	Parking and Secretariat Area	0.2	0.2
Amount of Land converted		**3**	**3**
Total Land Use		**100**	**100**

Source: [28]

The condition of the availability of Green Open Space in Karangkulon Hamlet based on current data shows that there is a decrease in the area of green open space compared to fourteen years ago. The area of green open space in question is rice fields and forests. Even though it is seen that the decline is only small, namely 3%, it has caused various impacts. One of the impacts is that in 2012 and 2019, floods occurred during the rainy season and drought during the dry season [29], [30]. This can happen due to a lack of public awareness of the importance of green open space in their environment. In addition, because the hamlet is located at the bottom of a hill, it is necessary to strengthen the green open space.

4.5. Conditions Affecting the Transfer of Green Open Space Functions in Giriloyo Batik Village

By making the Giriloyo Batik Village as a tourism area, there will be changes or additions to space to support tourism activities. According to Statistical Data of Tourist Visits to Giriloyo Batik Village from 2017 to 2019, the number of visits by foreign tourists and domestic tourists to Giriloyo Batik Village experienced ups and downs but the average number continued to increase. As happened in 2019, the number of tourist visits increased by 7310 tourists. It can be seen from the graph of tourist visits (Graph 1) in 2017 amounting to 20,714, in 2018 as many as 20,830 and in 2019 as many as 28,140 tourists. The increase in the number of visits is felt when the visit from schools or universities is running, namely from June to October. It was recorded that at one time the number of visits at one time amounted to 600 tourists with 12 bus fleets [28]. With this large number, the Giriloyo Batik Village has experienced an increase in capacity, so that the area expands its area which takes up a lot of green open space.

Chart 1. Graph of Tourist Progress to Giriloyo Batik Village (2017-2019)

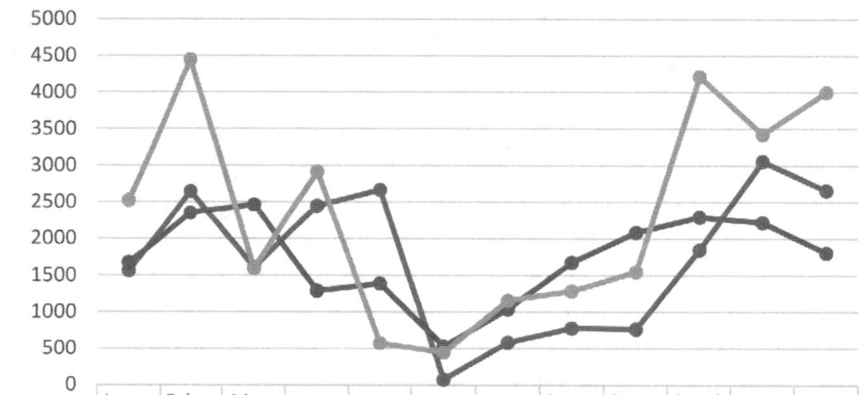

	January	February	March	April	May	June	July	August	September	October	November	December
2017	1565	2650	1608	2450	2669	73	581	780	768	1850	3059	2661
2018	1674	2352	2466	1290	1388	531	1039	1676	2088	2296	2224	1806
2019	2525	4451	1593	2915	576	445	1160	1284	1548	4217	3426	4000

Source: [28]

Green open space, which was originally a plantation or rice field, is now starting to change its function into built-up land which is used as land to support tourism. One of them is changing the function of rice fields, which is located 160 meters to the west of the Batik Gazebo, which is now used as a parking lot and secretariat (7). Because the Batik Gazebo is no longer able to accommodate the many tourists who visit, especially tourists who use tourism buses. In addition, the bustle of the Giriloyo Batik Village area has caused more and more residents to build new buildings around the Batik Gazebo. Residents who were originally already there began to add to the function of their homes by opening stalls or cottage industries.

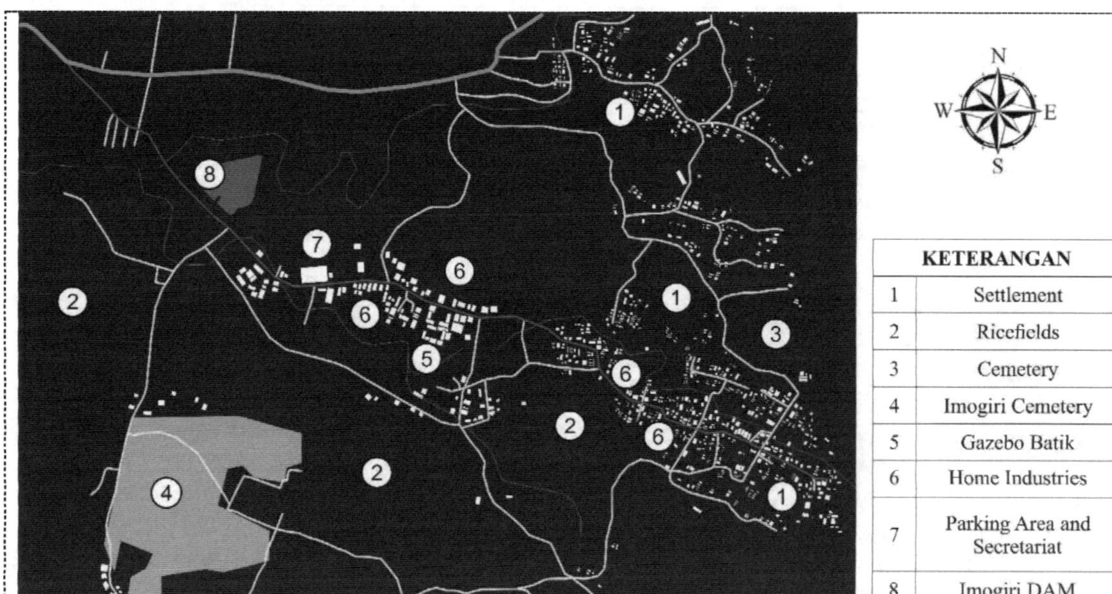

KETERANGAN	
1	Settlement
2	Ricefields
3	Cemetery
4	Imogiri Cemetery
5	Gazebo Batik
6	Home Industries
7	Parking Area and Secretariat
8	Imogiri DAM

Figure 3. Map of Land Use in Karangkulon Hamlet in 2021

Source: Google maps and information obtained from two informants

Karangkulon hamlet is a hamlet that is used as the center of the growth of batik tourism, where tourism activities develop. The development started from the establishment of the Batik Gazebo which was used as the gateway to the Giriloyo batik tourism area, in the Gazebo tourists could carry out various tourism activities ranging from learning batik to batik shopping tours. The movement of buildings began to mushroom around the Batik Gazebo, including parking areas, batik cottage industries, homestays, gurah medicine houses, and also grocery stores. With the development of existing supporting facilities, the existing green open space is decreasing. Green Open Space or green open space whose numbers are running low will cause various impacts, such as floods, droughts, or economic downturns for local residents.

4.6. Impact of Regional Development on the Availability of Green Open Space

The development of the area in Karangkulon Hamlet in the last fourteen years shows that an area of 3 hectares of green open space has been converted into built-up land. The built area includes a Batik Gazebo covering an area of 0.3 Ha, a parking and secretariat area of 0.2 Ha, and Embung Imogiri covering an area of 2.5 Ha. This cannot be separated from the development of Giriloyo Batik Village in the village. Giriloyo Batik Village is an edu-tourism destination located in Karangkulon Hamlet, Wukirsari Village, Imogiri District, Bantul Regency, Yogyakarta Special Region which has the potential for learning batik tourism, especially written batik.

Figure 4. Green Open Space Replacement Building
Source: Authors' documentation, 2021

The land that was originally open land is now busy being closed with built-up land as shown in Figure 4, the location of which is shown in points 5, 6, 7, and 8 in Figure 3. This problem has resulted in a decline in environmental quality where the existing green open space is increasingly narrowing. it will cause various environmental problems. In addition, there has also been a decline in the community's economy, where people who originally made a living as teak leaf laborers or farmers are now starting to shift. According to population data based on occupations in 2013, most of the residents make a living as agricultural or plantation laborers, amounting to 2532 people. Whereas in 2019 the number decreased to 1128 people, this happened one of them because their livelihoods began to shift [31].

IOP Conf. Series: Earth and Environmental Science **903** (2021) 012011 doi:10.1088/1755-1315/903/1/012011

4.7. Efforts to Overcome the Increased Utilization of Green Open Space as Built Space

Green open space should be maintained in order to maintain the quality of a comfortable, fresh, beautiful and clean-living environment. One of them is by means of collaboration between the community and the government so that later the benefits of the existence of green open spaces can be felt by the community. Such as the countermeasures that have been carried out by the government, namely the utilization and control of spatial planning. In Vilhem Lundsted's Welfare State concept regarding the view of social welfare, efforts to overcome environmental damage are by means of long-term inter-regional development. The goal is to provide a decent life in improving people's welfare [32].

5. Conclusion

From this research, it can be concluded that the development of the Karangkulon Hamlet area as a result of the construction of supporting facilities for the Giriloyo Batik Village tourism area in the last fourteen years has changed by 3%. These changes include the construction of tourism supporting facilities, namely the Batik Gazebo, parking and secretariat areas, as well as home industries (written batik, basic necessities, gurah). With the development of a tourism area, of course, it will also be accompanied by the development of tourism supporting facilities. As happened in the Giriloyo Batik Village, the change in the use of green open space as a support for tourism areas puts pressure on land in the form of converting green open spaces. So there is a need for an arrangement that is oriented to the preservation of the existence of green open space. One of the ways is the need for a clear legal order in strengthening the existence of green open spaces.

In this study, the relationship between green open space and urban development patterns has an impact on the reality that further studies are needed. Where the spatial planning of the Giriloyo Batik Village area is determined and influenced by changes in linkages. These changes are influenced by aspects of space utilization activities in the Batik Gazebo. Spatial planning in Giriloyo Batik Village is a deliberate change by the surrounding community in changing the pattern and structure of space utilization from one balance to a better balance.

6. Reference

[1] "Undang-Undang Tentang Penataan Ruang," Jakarta, 2007.

[2] H. Shirvani, The Urban Design Process, New York: Van Nostrand Reinhold Company, 1985.

[3] N. Purnomohadi, Ruang Terbuka Hijau Sebagai Unsur Utama Tata Ruang Kota, Jakarta: Direktorat Jendral Penataan Ruang, 2006.

[4] G. A. M. Suartika and A. R. Cuthbert, "The Sustainable Imperative-Smart Cities, Technology and Development," *Sustainability Journal Vol. 12. 8892,* vol. 12, pp. 1-15, 2020.

[5] A. R. Cuthbert and G. A. M. Suartika, "The Theatre of the Universe-Culture and Urban Design," *Journal of Urban Design,* vol. 19. No. 3, pp. 275-277, 2014.

[6] G. A. M. Suartika, "Perencanaan keruangan, wujud komunikasi kepentingan dalam pembangunan," *Jurnal Natah,* Vols. 5, No. 3, pp. 71-78, 2014.

[7] G. A. M. Suartika, "Cultural dimension of space and state urban planning practices: the conflicting interests in land development," in *Proceedings for the International Conference and Workshop of Urban Culture, Arte-Polis I*, Bandung, 2006.

[8] G. A. M. Suartika, "Pariwisata, masyarakat etnik dan proses komodifikasi," in *Proseding Seminar Perencanaan dan Pembangunan Spasial: Pariwisata Etnik*, Denpasar, Bali, 2010.

[9] J. Muchran, W. Ilham, M. Siddiq and Susilawati, "Model Perencanaan Ruang Terbuka Hijau Taman Lingkungan di Kota Banjarbaru, Kalimantan Selatan," *EnviroScienteae (11),* pp. 19-32, 2015.

[10] Samsudi, "Ruang Terbuka Hijau Kebutuhan Tata Ruang Perkotaan Kota Surakarta," *Journal of Rural and Development Vol. 1, No. 1,* pp. 11-19, 2010.

[11] A. Dwiyanto, "Kuantitas dan Kualitas Ruang Terbuka Hijau di Permukiman Perkotaan," *TEKNIK - Vol. 3 N. 2,* pp. 88-93, 2009.

[12] T. Angelia, "Konsep Pengembangan Ruang Terbuka Hijau Sebagai Fungsi Ekologis Penyerap Air Hujan di Kecamatan Rungkut Kota Surabaya," Institut Teknologi Sepuluh Nopember, Surabaya, 2017.

[13] L. Muliawati, "Perencanaan Lanskap Wisata Dukuh Karangkulon di Desa Batik Wukirsari Imogiri Yogyakarta," Institut Pertanian Bogor, Bogor, 2012.

[14] R. T. Wahyuhana and M. Muktiali, "Pengaruh Kawasan Wisata Sendang Asri Waduk Gajah Mungkur Terhadap Perubahan Guna Lahan dan Aspek Sosial-Ekonomi Masyarakat," *Jurnal Teknik PWK Vol. 2 (3),* pp. 500-509, 2013.

[15] I. W. T. Sidarta, "Dampak Perkembangan Pariwisata Terhadap Kondisi Lingkungan, Sosial, dan Ekonomi Masyarakat (Studi Kasus Kawasan Pariwisata Sanur, Denpasar-Bali)," Universitas Diponegoro, Semarang, 2002.

[16] R. Trancik, Finding Lost Spaces: Theories of Urban Design, USA: John Wiiley and Sons, 1986.

[17] I. Arianti, "Ruang Terbuka Hijau," *Jurnal Ilmu Pengetahuan dan Rekayasa,* pp. 1-7, 2010.

[18] S. M. Lesil, "Pontianak Waterfront City Sebagai Obyek Wisata Ruang Terbuka Publik," Universitas Atma Jaya Yogyakarta, Yogyakarta, 2016.

[19] I. M. Y. Prasada and T. A. Rosa, "Dampak Alih Funsi Lahan Sawah Terhadap Ketahanan Pangan di Daerah Istimewa Yogyakarta," *Jurnal Sosial Ekonomi Pertanian Vo. 14, (3),* pp. 210-224, 2018.

[20] Badan Pusat Statistik (BPS), "Luas Lahan Menurut Penggunaan," Daerah Istimewa Yogyakarta, 2016.

[21] R. Hidayah, Sativa and S. H, "Strategi Pemenuhan Ruang Terbuka Hijau Publik di Kota Yogyakarta," *INERSIA, Vol. 17(1),* pp. 11-18, 2021.

[22] K. D. Indarto and S. Rahayu, "Dampak Pembangunan Perumahan Terhadap Kondisi Lingkungan, Sosial dan Ekonomi Masyarakat Sekitar di Kelurahan Sambiroto, Kecamatan Tembalang," *Jurnal Teknik PWK, Vol. 4 (3),* pp. 428-439, 2015.

[23] I. G. A. D. Muliasari and Widiastuti, "Daya Dukung Lingkungan Terkait Pengolahan Limbah Batik di Kampung Batik Giriloyo, Kabupaten Bantul, Yogyakarta," *ATRIUM, Vol. 6(2),* pp. 131-139, 2020.

[24] R. A. Sari, "Kajian Perkembangan Kota Batang Berdasarkan Struktur Ruang Kota," Universitas Diponegoro, Semarang, 2008.

[25] Hardani, H. A. Nur, A. Helmina, A. F. Roushandy, U. Jumari, F. U. Evi, J. S. Dhika and R. I. Ria, Metode Penelitian Kualitatif & Kuantitatif, Yogyakarta: CV. Pustaka Ilmu Group, 2020.

[26] I. G. A. D. Muliasari, "Pusat Informasi Pariwisata dan Amenitas Kampung Batik Giriloyo, Dusun Karangkulon, Desa Wukirsari, Kecamatan Imogiri, Kabupaten Bantul," Universitas Atma Jaya Yogyakarta, Yogyakarta, 2019.

[27] Kampung Batik Giriloyo, "Histori Kampung Batik Giriloyo," Batik Giriloyo, [Online]. Available: https://batikgiriloyo.com/histori/. [Accessed 2019].

[28] Paguyuban Batik Tulis Giriloyo, "Pengelolaan Kampung Batik Giriloyo," Kampung Batik Giriloyo, Yogyakarta, 2019.

[29] Y. H., "Hujan Seharian 9 Kecamatan di Bantul Terendam Banjir," 18 March 2019. [Online]. Available: https://www.liputan6.com/regional/read/3919343/hujan-seharian-9-kecamatan-di-bantul-terendam-banjir. [Accessed 10 September 2021].

[30] Satu Data Bantul, "Kejadian Bencana Banjir per Kecamatan," [Online]. Available: https://data.bantulkab.go.id/tr/dataset/bencana-banjir/. [Accessed 27 September 2021].

[31] Kelurahan Wukirsari, "wukirsari.bantulkab," 24 September 2021. [Online]. Available: https://wukirsari.bantulkab.go.id/first/penduduk/pekerjaan.

[32] D. Riskanita and Y. Widowaty, "Upaya Pemerintah Daerah Mengatasi Kerusakan Lingkungan Akibat Alih Fungsi Lahan Berdasarkan Konsep Negara Kesejahteraan," *Jurnal Penelitian Hukum, Vol. 28 (2),* pp. 123-134, 2019.

7. Acknowledgement

Authors' gratitude is dedicated to:

1. The manager of Giriloyo Batik Association, Wukirsari Village, Imogiri District, Bantul Regency, Yogyakarta Special Region for the time, opportunity, and assistance during the conduct of this research.
2. The University of Udayana-Bali, Indonesia for enabling the publication of this article by providing research grants (with a contract No of: B/96-111/UN14.4.A/PT.01.05/2021, dated 3 Mei 2021).

Road Intersection as a Public Space and Venue for Advertisements in an Urban Setting

I Wayan Yoga Bahagia Putra,[1] Gusti Ayu Made Suartika[1] and Kadek Edi Saputra[1]

[1] Master Program in Architecture, Development and PLanning, Udayana University – Bali, Indonesia

Email: yogabahagiaputra@gmail.com

Abstrak. Advertising media uses a lot of public space as a place to advertise which can reduce the function of public space in urban areas. This study aims to determine the use of urban space as a medium for outdoor advertising that interrupts and reduces the function of space. This study uses a qualitative approach with descriptive data analysis methods. The four intersections on *Jalan Raya Dalung* have interrupted the function of public spaces, including functions as pedestrians (pedestrian paths), *telajakan* (as an element of reforestation), protecting the environment, and as borders. Inappropriate placement of advertising media creates a chaotic impression and visual chaos.

Keywords: advertising media, public spaces, crossroads

1. Introduction

The rapid development of the city raises a number of trends in the use of spaces within the city as the face of advertising media. Outdoor advertising media in urban areas such as billboards, neon box banners, video trons and posters continue to decorate corners of the city with various commercial, public, and informational purposes. Advertising activities through digital media have not been able to completely replace conventional outdoor advertising. Advertising conventionally such as installing billboards effectively has a reach (penetration) of 54 percent of the contents of the heads of people who pass the billboards [1].

The installation of advertising media in urban areas generally utilizes space that is widely traversed and seen by the public. Crossroads are one of the strategic places that are widely used to place advertising media by the public, private sector and government. We know that crossroads are public open spaces, spaces where community mobility occurs. Crossroads also consist of pedestrian paths, signposts, city parks, guard posts, rest areas, statues, monuments, and gates [2]. Advertising media installed at crossroads still cause a lot of disturbance to road users, discomfort, and the absence of standards for installing advertising media will affect the beauty of an environment [3].

Public open space is an important element in maintaining the efficient operation of life in urban areas whose existence and quality need to be maintained [27, 28, 29, 30]. Public open space is a space that belongs to all citizens of the city that reflects the character and has the function of social interaction for the people of the city of Kussoy [4]. Public open space is also expected to accommodate all activities and provide comfort for its users. Public open spaces should be well designed to maintain environmental balance between built and unbuilt spaces. According to Stephen Cars [5] in his book, Public Spaces, public spaces in the form of parks, environmental parks, pedestrian roads, crossroads, parking lots, plazas, all of which are public open spaces of the city that accommodate various city activities and places for social interaction of urban communities.

I Gusti Ngurah Bagus Kusuma Putra in 2018 [6] has conducted a study on the phenomenon of installing outdoor advertising media at several intersections in Denpasar City. This study discusses regulations and violations of advertisements that are not culturally aesthetic and shows the level of satisfaction of respondents with promotional media in the city of Denpasar. Then in 2016 [7], Christianna conducted research on signage design as a solution to visual pollution in urban areas. Inappropriately placed advertising media often causes visual chaos on the road [8]. Roads in major cities in the world will not be able to escape the siege of signage in the form of signage, billboards, billboards and others. This indirectly also causes problems in the city related to many violations in the installation of signage that cause visual pollution. Signage that is not neatly arranged first will reduce the aesthetics of the city, the two cities will lose their uniqueness, thirdly the installation of excessive signage in public spaces will cause medical or psychological disorders such as stress, headaches, and aggressiveness and fourth visual pollution will be able to disrupt the concentration of motorists. The signage design by aligning local wisdom is expected to be a solution to this visual pollution. The existence of good coordination between architecture and urban planning will greatly reduce the clutter of signage and advertising media in a city.

Badung experienced a similar situation in the placement of advertising media in public open spaces. Many advertising media are placed in inappropriate positions, as is often seen at several intersections. Improper placement will disrupt the function of public spaces at crossroads. The research that will be conducted at the intersection of *Jalan Raya Dalung* aims to determine the use of urban space as an advertising media platform that interrupts and reduces the function of public spaces in urban areas.

2. Public Space and Urban Liveability

According to Darmawan [9] Public space is a space belonging to the people, rich or poor, young and old, men and women with all their social activities. City public spaces that are categorized as "good" must be able to accommodate the activities of their stakeholders fairly without discrimination, safe, comfortable, easy and freely accessible [5]. In public spaces, citizens can meet, interact in their spare time and fully actualize themselves where social roles at home and in the office are generally limited [10].

The road as one of the linear public open spaces [11], is an element of urban space that is very important in supporting the needs of citizens. Roads have a function and role as a medium of movement, a medium for the exchange of goods, services, and information (exchange) as well as a regulator of environmental and ecological conditions.[12]. As one of the public spaces, roads and road sections in the form of intersections are expected to become public facilities that can be used by city residents.

One of the studies conducted by Untung Joko Cahyono in 2012 [2] regarding the management of public spaces around road junctions. This study explains that the intersection is a strategic place that has a very high conflict of interest. Many parties are attracted to this crossroads because at the crossroads people will stop, walk slowly, be alert and careful. The community, government, and private sector will compete to take advantage of this crossroads according to their respective interests. One example of the use of road crossings as public spaces discussed by Joko Cahyono, stated that the Hotel Indonesia Roundabout (HI Roundabout) as a public space at a crossroads that functions well can accommodate the function of public spaces. The roundabout has a traffic function, namely the meeting of Jalan M.H Thamrin which intersects with *Jalan Imambonjol, Jalan Sutan Syahrir* and *Jalan Kebon Kacang*. HI

Roundabout is also known as a landmark of the City of Jakarta. In its social use, the HI roundabout is often used by residents for demonstrations, Car Free Day (CFD) activities and cultural performances.

Based on the explanation above, it can be concluded that the intersection as a public open space has an important role for all social activities of the city community. Public space at the crossroads is expected to be a space for interaction that accommodates the interests of various parties from the community, private sector, and government in a fair, safe, and comfortable manner without discrimination.

3. Public Space as a Media for Advertising

Public space with the development of the city in the modern era is widely used in various ways by city residents, the private sector and the government. One of its uses that is rife in urban areas is as a place for placing outside advertising media. In a good city design, public spaces that are used as media for advertising must pay attention to the visual area, namely the ease of a citizen to identify a visual area that is harmonious, proportional and aesthetic [13]. The fact that occurs in the field, there are still many installations of outdoor advertising media that have not paid attention to the aesthetic value of an area

Edi Purwanto in 2014 conducted a study on the privatization of public space from a civic center to a central business district in the Simpang Lima area of Semarang [14]. One of these studies discussed advertising media in the Simpang Lima area of Semarang. The presence of advertising media at the intersection tends to take advantage of the economic potential of the region rather than viewing the area from its aesthetic value. The form of the proportion and scale of advertising media in the area is not considered and is very disturbing to public spaces. The advertising media installed are dominated by billboards, billboards, and banners (commercial identity).

The installation of advertising media in the Simpang Lima area of Semarang, apart from on land, also occurs in public spaces that are in the air. The expectations of city residents to get a good, beautiful and comfortable visual view of public spaces are not fulfilled. Citizens must be faced with the sight of advertising media placed by business people, and the government with various business objectives and reasons.

Figure 1. Advertisements at Simpang Lima Semarang
Source: [14]

Based on the explanation above, it can be concluded that public spaces used as media for advertising must pay attention to the visual value, beauty and aesthetics of the area. Advertising media installed at intersections so as not to interfere with the function of the area as a public space and the needs of citizens for public space.

IOP Conf. Series: Earth and Environmental Science **903** (2021) 012012 doi:10.1088/1755-1315/903/1/012012

4. Maintaining the Existence of Public Spaces

The spatial plan of a city is one way to solve the chaotic problem of installing advertising media which sometimes interferes with the functions of other public spaces in urban areas. Outdoor advertising media are media that are installed in open places such as roadsides, city centers, city buses, fences and walls [15]. The arrangement of outdoor advertising media can help build a sense of place in an urban area [16]. To achieve a sense of place in an area, the arrangement of outdoor advertising media is very necessary to create order and make the road more lively [17] .

Prayanto Widyo Harsanto [18] also studied a similar problem that occurred in Yogyakarta. The streets in Yogyakarta, which are crowded with the activities of road users, experience confusion in the placement of advertising media. Functionally, advertising media is made to market a product, provide information, and various other purposes. Advertising media for the government is a source of regional income. There must be a good arrangement according to the city's spatial layout so that the advertising media is more pleasing to the eye, does not damage the view and violates the rules. This arrangement is also expected to restore Yogyakarta's slogan "Jogja is special and Jogja has a comfortable heart". The slogan indicates that comfort is needed in the design of a city's public spaces.

Based on the explanation above, it can be concluded that urban spatial planning and regional regulations are needed to bridge and accommodate the procedures for placing advertising media in public open spaces. Advertising media in a city is something that cannot be avoided and must still exist to liven up the atmosphere of the city. Basically, public open space as a place for advertising in urban areas must be maintained to create a city that is comfortable and harmonious with all the activities of its citizens

5. Rules Governing the Accommodation of Advertising Medias in Public Spaces

Until now, Badung Regency already has regulations regarding the installation of billboards, namely Regional Regulation (Perda) Number 80, Year 2014 [19] About the Implementation of Advertisements. Article 1 paragraph 10 of this regulation regulates the installation of billboards measuring more than 10 m2. The pattern of distributing advertisements must refer to the Master Plan by considering norms, religion, norms of decency, morality, health, public order and security. Areas in Badung Regency that are of concern to the government in placing advertising media are parks that become regional mascots/icons, around the Ngurah Rai Statue, underpass areas, along the coast, the *Catus Pata* area of *Mengwi Village* and *Taman Ayun*, the entrance area to the Regional Government Center, parks assessment, intersection and median facilitated by the government/local government (figure 2).

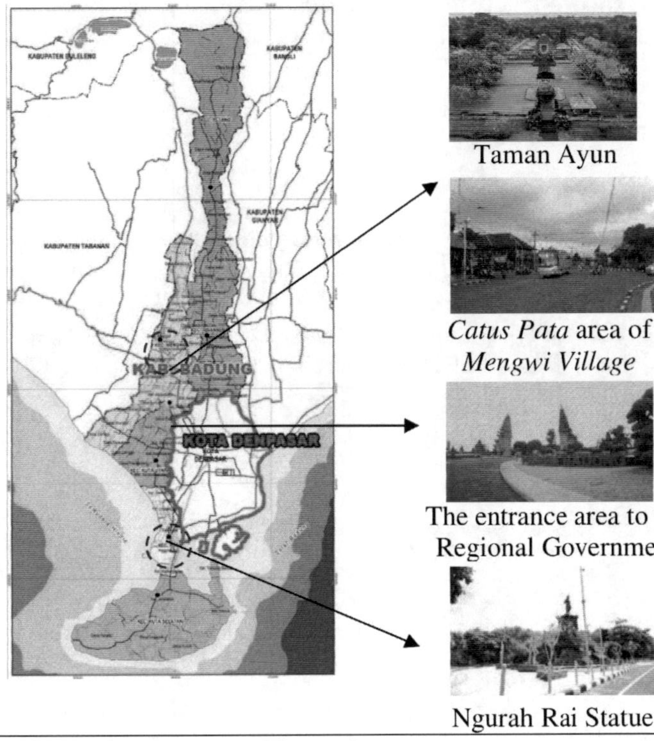

Figure 2.
Map of Areas of Concern for Placement of Advertising Media in Badung
Source*: Google Accessed on 2 August 2021*

Advertising ethics that regulate outdoor media in advertising are as follows [20] :

1. Installed in a location that has obtained permission from the authorities
2. Obligation to respect and protect the building or the environment and preserve the monument
3. Outdoor advertisements may not be placed over other advertisements.
4. Cannot be installed close to competing products
5. Foundations, structures and panels for outdoor advertisements in the form of billboards must go through civil and mechanical calculations to ensure the safety and comfort of the surrounding community.
6. The construction and advertising area must be physically and aesthetically harmonious with the building and its surrounding environment in accordance with applicable local regulations
7. Outdoor advertisements in the form of billboards are not placed on medians, separators, or road islands
8. Outdoor advertising must not cover traffic signs, traffic signs and markings
9. Outdoor advertising lighting arrangements should not dazzle road users
10. Ads about liquor may only be placed in areas designated for an adult audience

6. Method

The research uses a qualitative descriptive method with a case study approach. The selection of descriptive methods is expected to describe the condition of public spaces used for advertising media at several intersections in Badung Regency. Data was collected by observation using research instruments

IOP Conf. Series: Earth and Environmental Science **903** (2021) 012012 doi:10.1088/1755-1315/903/1/012012

in the form of cameras, and stationery to support research data. Literature studies are carried out through literature searches including books, journals, internet data and references related to the research to be carried out.

7. Data, Discussion, and Results

7.1 Location Overview

The intersection locations studied are Kwanji Intersection with map coordinates 8°36'36.3" South Latitude (LS), 115°10'41.9" East Longitude (BT), Bangsal Intersection with coordinates 8°36'39.0" South Latitude, 115°10'33.4" East Longitude, Dalung Village Office Intersection 8°36'38.7" South Latitude, 115°10'20.1" East Longitude and Toko Iwaka Dalung Interchange -8°36'24.8" South Latitude, 115°10'17.5 East Longitude. heavy traffic and there is outdoor advertising media.

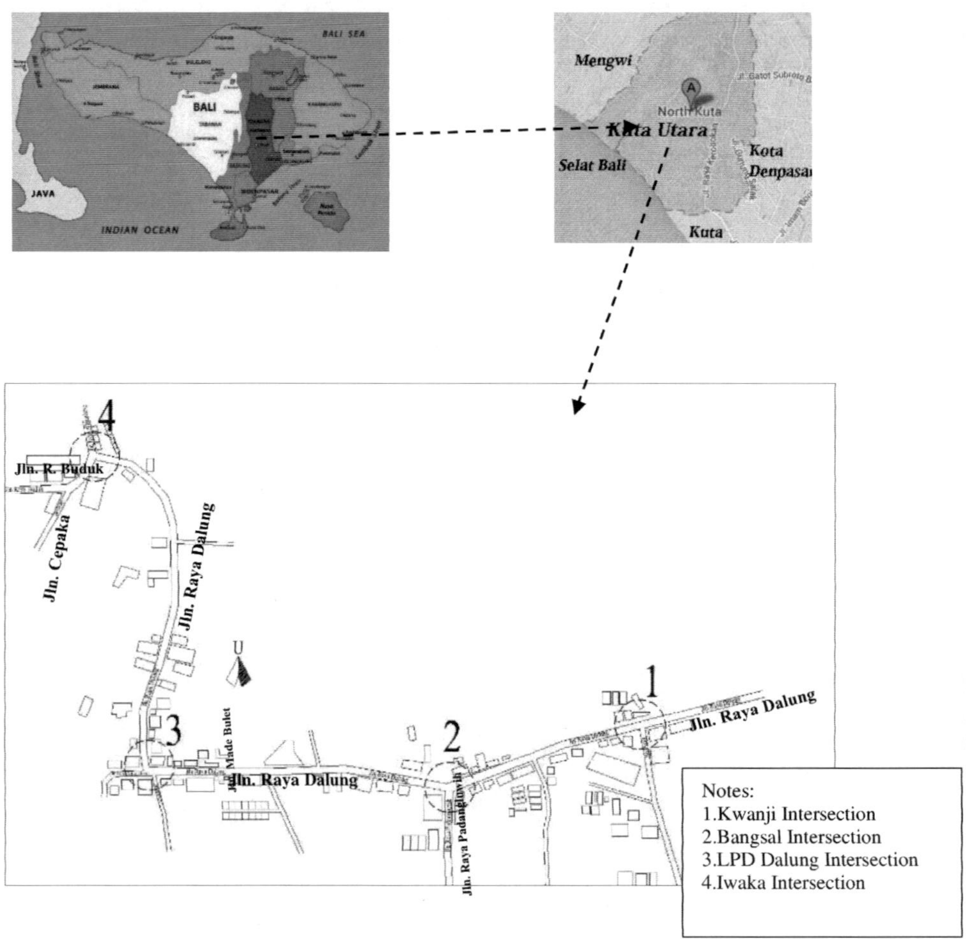

Figure 3 . Intersection Location on Jalan Raya Dalung
Source*: Google Maps accessed on 2 August 2021*

IOP Conf. Series: Earth and Environmental Science **903** (2021) 012012 doi:10.1088/1755-1315/903/1/012012

7.2 Kwanji Intersection

Kwanji Intersection has a high traffic density and is classified as a small intersection. The condition of the intersection often experiences traffic jams during working hours and coming home from work [21] (balipuspanews, 5 Dec 2018). The intersection has elements of public open space in the form of roads, pedestrians that are quite well maintained and there are several advertising media at this intersection in the form of billboards, banners, and pamphlets. The width of the road is approximately 5.5 meters with pedestrian paths / sidewalks on both sides of the road. The intersection is also surrounded by houses and shops with trading and residential activities. The intersection is a meeting place between Jalan Raya Dalung and Jalan Panji which does not yet have a special space for placing advertising media. The placement of advertising media utilizes sidewalks and traffic signs. The ads that are installed are temporary ads that seem chaotic without a good design with various dimensions.

7.2.1 Intersection-as a Public Space, Its Functions, Its Uses as Advertising Venues, and Associated Problems

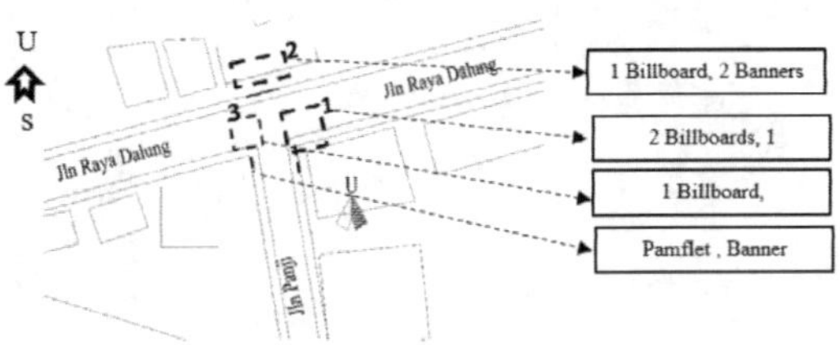

Figure 4. Kwanji Intersection Location Map

Figure 5. Kwanji Junction Conditions
Source*: Authors*

Kwanji intersection is a T-junction that has elements of public space in the form of roads, pedestrians, and street furniture. This intersection has a traffic function as a motorized vehicle circulation space, aesthetics as the beauty of the city corner and pedestrian circulation. The placement of advertising

IOP Conf. Series: Earth and Environmental Science **903** (2021) 012012 doi:10.1088/1755-1315/903/1/012012

media at Kwanji Intersection is mostly placed on pedestrian paths and slopes with advertising orientations to the west (figure 5, item 1), south (figure 5, item 2) and east (figure 5, item3).

Pedestrian paths are very important in urban areas as amenities that function as a safe and comfortable place for pedestrians [22]. Pedestrian paths are very much needed by city residents to move from one place to another easily, safely and comfortably.

The pedestrian path at the Kwanji intersection based on observations is designed for pedestrian activities with paving material with a width of approximately 1.2 meters. The use of pedestrians as an advertising platform that takes up approximately 60 centimetres of pedestrian path space based on observations will interfere with the safety and comfort of pedestrians. Pedestrians are also disturbed because of the presence of stones as a barrier to the construction of billboards that cover the pedestrians.

Figure 6. Kwanji Intersection Pedestrian Trail
Source*: Author*

In addition, the visual comfort of motorists from the south, east and west is also disturbed by the placement of advertising media. The placement of advertising media on three sides of the road also reduces the beauty of the corner of the city and disturbs the elements of public space at the intersection

7.3 Bangsal Dalung Intersection

Bangsal intersection is an intersection with high traffic activity. The width of the road is 5.5 meters with pedestrian paths on both sides of the road. The character of the intersection is very closely related to the history of the Bangsal Struggle Monument. The struggle monument is a very important historical building and is protected by the state [23]. This intersection has a traffic function as a motorized vehicle circulation space, aesthetics as the beauty of the city corner and pedestrian circulation. The social function is as a communication space between pedestrians, and the location of the monument to the ward struggle. This location is also surrounded by shophouse buildings with trading activities, ward monument tours, and parks on the north side of the slope.

7.3.1 Intersection-as a Public Space, Its Functions, Its Uses as Advertising Venues, and Associated Problems

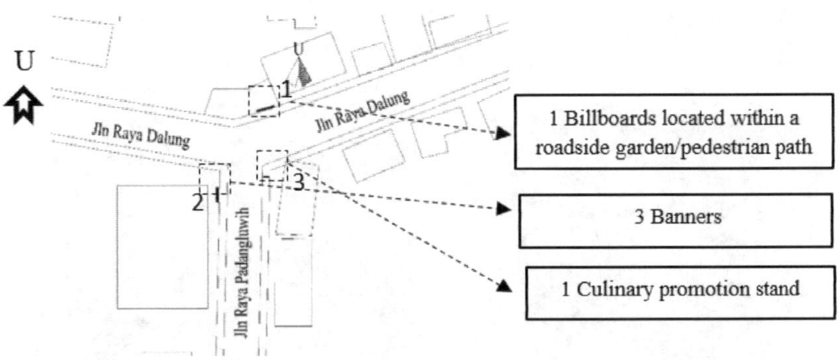

Figure 7. Intersection Location Map, Bangsal Dalung
Source*: Authors*

Figure 8. Intersection Condition in Bangsal
Source*: Authors*

Bangsal intersection is a T-junction with high traffic density. The advertising media installed at the Bangsal Intersection are three banners oriented to the east (figure 8 item 2), one food promotion stand on the pedestrian street (picture 8 item 3), and one billboard above the park with an orientation to the south (picture 8). , item 1). This type of advertising media is temporary. The placement of the advertising media utilizes pedestrians and gardens at the slopes.

Telajakan is one of the public spaces in urban areas which is interrupted by the placement of advertising media. Physically, Telajakan has the first three functions as reforestation, protecting the environment, and maintaining building boundaries for the safety and comfort of the building and its surrounding environment [24]. The placement of advertising media was deemed inappropriate and interrupted the tread function at the intersection of the wards which were used as parks. The park at the intersection of the wards is very closely related to the values and concepts of belief related to the ward struggle. Telajakan in the form of a garden in front of a house in Bali is very close to the meaning of aesthetics, security, social and spiritual meaning [25]

Figure 9. Condition of the Walk/Garden at the Bangsal Junction
Source: Authors

The placement of advertising media is also placed at the pedestrian crossing of the ward. Every pedestrian in the city is expected to be a space for movement where data provides a sense of security and comfort. Pedestrians placed on temporary food promotion stands will disturb the space for pedestrians. The pedestrian path which has a width of 1.2 meters will be disturbed in accommodating pedestrian activities because half the width of the pedestrian is taken by advertising media (Figure 10).

Figure 10. The Condition of the Bangsal Crossing Pedestrian
Source: *Author*

7.4 LPD Intersection Dalung

LPD Dalung intersection is an intersection which is a traffic jam-prone point [26]. Congestion most often occurs during working hours and at work hours. The LPD Dalung intersection is classified as a small intersection with a road width of 5.5 meters with pedestrian paths on both sides of the road.

The character of the intersection is an office and shop area. The intersection is very dense with motorized vehicle activity, and the location of this intersection is close to schools. The social function is as a communication space between pedestrians. The traffic function of this intersection is a meeting room between Jalan Padang Luwih, Jalan Raya Dalung, Jalan Made Bulet and Jalan Gusti Ngurah Gentuh.

7.4.1 Intersection-as a Public Space, Its Functions, Its Uses as Advertising Venues, and Associated Problems

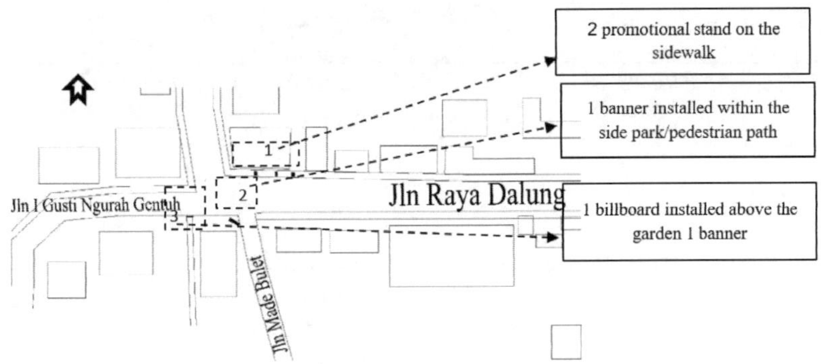

Figure 11. Map Location of Intersection LPD Dalung
Source*: Authors*

Figure 12. Kondition of Intersection of LPD Dalung
Source*: Authors*

The LPD Dalung intersection is an intersection. The advertising media at the Dalung LPD Intersection are in the form of a promotional stand on the sidewalk with an east-west orientation (Figure 12, item 1) Banners above the park or an easterly orientated walkway (figure 12, item 2) and billboards on top of the walkway, banners with orientation north. (Picture 12, item 3). The placement of the advertising media on the pedestrian area interferes with pedestrian circulation, on top of the park/walk, disturbs the beauty of the corner of the city. One of the conditions of the park at the intersection of LPD Dalung where advertising media is placed. The condition of the garden is not maintained and the advertising media seems to cover the angkuls and the advertising foundation seems to damage the plants as seen in picture 13.

Figure 13. Junction Park Condition of LPD Dalung
Source: Authors

7.5 Iwaka Store Junction

Toko Iwaka intersection has a road width of 5.5 meters with pedestrian paths on both sides of the road with a width of 1.2 meters. Elements of public space at the intersection in the form of roads, pedestrians, and parks / trails. The character of the intersection is a trading area. The intersection is very dense with motorized vehicles, and the location of this intersection is close to a traditional market. Social functions as a place of communication include pedestrians, jogging, and selling. The intersection is surrounded by shop houses with trading activities. Traffic functions as a meeting place between the Abianbase, Buduk, and Cepaka roads.

7.5.1 Intersection-as a Public Space, Its Functions, Its Uses as Advertising Venues, and Associated Problems

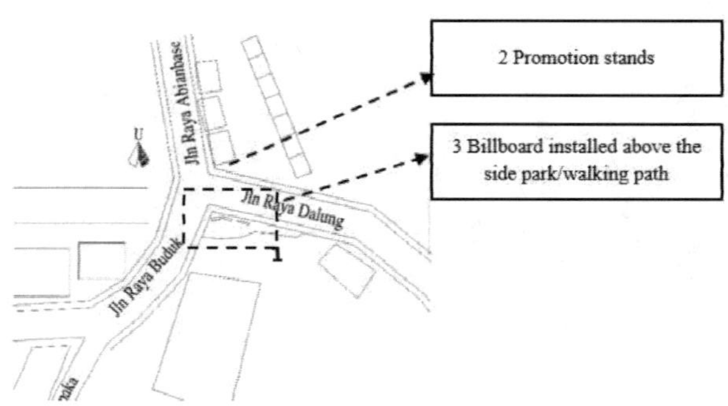

Figure 14. Intersection Location Map of Iwaka
Source: Authors

Figure 15. Kondition Intersection of Iwaka
Source: Authors

The Iwaka intersection is in the form of a warning with dense traffic conditions. There are three advertising media at Iwaka Junction, billboards, and a promotional booth. Billboards are placed in permanent parks/walks that are commercially rented. The placement of the three billboards interrupts the climbing function which functions as greenery. The presence of advertising media on the slopes disturbs the space for the growth of trees. The orientation of the billboard to the north interferes with the view and circulation of vehicles from the north.

Advertising media in the form of promotional stands are mostly placed on pedestrians which interfere with pedestrian activities. Pedestrians as pedestrian spaces cannot function optimally because the placement of promotional stands covers almost all pedestrian access as seen in Figure 16.

Figure 16. Crossroad Pedestrian Condition of Iwaka
Source: Authors

8. Conclusion

The placement of advertising media at the four intersections of Jalan Raya Dalung largely interrupts the function of public spaces in the four research locations. The function of the public space that is interrupted is pedestrian with the main function of the pedestrian path, treads with the main function of

IOP Conf. Series: Earth and Environmental Science **903** (2021) 012012 doi:10.1088/1755-1315/903/1/012012

reforestation, protecting the environment, and maintaining building boundaries for the safety and comfort of the building and its surrounding environment. By using the space as an outdoor advertising media container in the four research locations, the pedestrian function used for pedestrian activities is now not optimal because half the width is used to place advertising media at several intersection points. Gardens with aesthetic functions are also declining due to the placement of advertising media that cover plants. The reforestation function has also decreased due to the placement of high advertising media and disturbing the space for plant growth in a relatively narrow garden. The placement of advertising media that is not in accordance with the rules will create a chaotic impression and visual chaos. The author hopes that the government needs to make strict rules [31, 32, 33, 34] and clear mapping regarding the installation of advertising medias in public spaces.

9. References

[1] Kompas.com, "Puan di Baliho, Ganjar di Media Sosial," 2021.
[2] U. J. Cahyono, 2012 "Pengelolaan Ruang Publik di Sekitar Persimpangan Jalan," vol. 10 no 2, pp. 44-54, 2012.
[3] A. Murtomo, 2007 "Penataan Papan Reklame pada Penggal Jalan Hayam Wuruk Semarang," .
[4] R. Z. Sentono and dkk, 2014 Ketersedian dan Kebutuhan Ruang Terbuka Publik Kelurahan Bitung Karang Ria di Kota Manado, vol. 6 no 1
[5] S. Cars and dkk 1992 , Public Spaces, Combridge University Press
[6] G. N. B. K. P. Putra and G. A. M. Suartika, 2018 "Fenomena Media Periklanan Terhadap Citra Kota Denpasar Studi Kasus di Beberapa Persimpangan Kota Denpasar," vol. 5
[7] A. Christianna, "Desain Signage Sebagai Solusi Pencemaran Visual," Prosiding Seminar Nasional Lingkungan Hidup. Departemen Matakuliah Umum Universitas Kristen Petra, 2016.
[8] P. D. Spreiregen, "Design Competitions," 1979.
[9] E. Darmawan,2005 "Ruang Publik dan Kualitas Ruang Kota," Prosiding Seminar Nasional Pesat,
[10] P. Hariyono, 2007. Sosiologi Kota untuk Arsitek, Bumi Aksara Jakarta, 2007.
[11] A. Rossi, 1982. "Street Arsitekture in City Problem and Operations Case Study Jalan Sudirman Jogjakarta,"
[12] Ikaputra, 2001 "Street Architecture In City Problem and Operations. Case Study Jalan Sudirman Yogyakarta," Makalah ini disampaikan pada Seminar Internasional Universitas Katolik Parahyangan Bandung,
[13] W. H. Whyte,1980. The Social Life of Smal Urban Spaces, The Concervation Foundation, Washington DC
[14] E. Purwanto,2014 "Privatisasi Ruang Publik dari Civic Centre Menjadi Central Business District (Belajar dari Kasus Kawasan Simpang Lima Semarang)," vol. Tata Loka Volume 16
[15] R. Khasali,1995 "Manajemen Periklanan,"
[16] H. Shirvani, 1895"Urban Design Process,"
[17] A. A. N. Aritama,2016 "Penanda Fungsi dan Media Promosi di Koridor Jalan Hayam Wuruk Denpasar," Space, vol. 3 No 3, pp. 276-290
[18] P. W. Harsanto, 2014 "Yogyakarta dalam Kepungan Iklan (Sebuah Kajian tentang Estetika Lingkungan)," Jurnal Dekade, vol. 7, pp. 45-55
[19] "Perda Bupati Badung Nomor 80 Tahun 2014 Tentang Penyelenggaraan Rekame di Kabuaten Badung,"
[20] "Dewan Periklanan Indonesia (2007) . Etika Pariwara Indonesia (Tata Krama dan Tata Cara Periklanan Indonesia). Jakarta,"
[21] "Balipuspanews.com 2018. Uji Coba Lampu Lalu Lintas di Simpang Kwanji Dalung".
[22] L. Mauliani,2010 "Fungsi dan Peran Jalur Pedestrian Sebagai bagi Pejalan Kaki Sebuah Studi Banding Terhadap Fungsi Pedestrian," Nalars Volume 9
[23] N. C. Setiaji and M. Hanif, 2018 "Kajian Makna Simbolis Patung dan Monumen di Kabupaten Ponorogo Sebagai Sumber Pembelajaran Sejarah Lokal," Jurnal Agastya , vol. 08 No 01, 2018.
[24] N. Dwijendra, 2010 "Arsitektur Rumah Tradisional Bali", Denpasar: Undayana University Press,

[25] N. A. Dwijendra,2013 ""Telajakan" Ruang Terbuka Hijau Tradisional Bali," Jurnal Ipteks "New Media" Volume 4 Nomor 2,
[26] TribunBali.com,2019 "Jalur di Badung Ini Bakal Macet Jelang Tahun Baru, Ini Jalan Alternatif yang Bisa Dilalui,"
[27] Suartika, G A M, Zerby, J, Cuthbert, A R, (2018) 'Doors of Perception to Space – Time - Meaning: Ideology, Religion, and Aesthetics in Balinese Development' *Journal Space and Culture,* Vol 21 (4) November, p: 340-357.
[28] Suartika,G A M (2006) 'Cultural dimension of space and state urban planning practices: the conflicting interests in land development,' *Proceedings for the International Conference and Workshop of Urban Culture, Arte-Polis 1,* Department of Architecture, Planning and Policy Development, Institute of Technology Bandung, 21-23 July 2006.
[29] Suartika, G A M, Mudra, I K, and Saputra, K E (2019) An Open Market: The Legitimation and Regulation of the Public Realm in Denpasar, in IOP Conf. Ser.: Earth Environ. Sci. 396 012041, p: 1-8.
[30] Swanendri, N M and Suartika, G A M (2019) Try and Find it - Understanding Public Space and Civil Society in Bali, in IOP Conf. Ser.: Earth Environ. Sci. 396 012030, p: 1-11.
[31] Suartika, G A M, Cuthbert, A R, (2020) ' Sleight of Hand: The Expropriation of Balinese Culture ' *Journal Space and Culture,* Online edition.
[32] Suartika, G A M and Saputra, K E (2019), 'The Creation of Harmonious Living Spheres - Landscape, Capital, and the Balinese Way' in IOP Conf. Series: Earth and Environmental Science 2019 IOP Conf. Ser.: Earth Environ. Sci. 248 012065, p: 1-12.
[33] Suartika, G A M (2019), 'Lessons from a Small Island – Density, Spatial Development, and Identity' in IOP Conf. Series: Earth and Environmental Science, 2019 IOP Conf. Ser.: Earth Environ. Sci. 248 012016, p: 1-13.
[34] Suartika, G A M (2013) 'Pembangunan Identitas Lokal dan Kepadatan Spasial: Studi Kasus Bali' *Proseding Seminar Nasional: Reinterpretasi Identitas Arsitektur Nusantara* Jurusan Arsitektur, Fakultas Teknik, Universitas Udayana, Gedung Pascasarjana Universitas Udayana, Denpasar, 10 Oktober 2013.

Acknowledgement

Authors are grateful to the Indonesian Ministerial for Education, Culture, Research and Technology for providing research grants (with a contract No of: B/136-10/UN14.4.A/PT.01.05/2021 and dated 29 June 2021) that enable the publication of this article.

IOP Conf. Series: Earth and Environmental Science **903** (2021) 012013 doi:10.1088/1755-1315/903/1/012013

Riverfront Development - A Public Realm, Its Adaptations, and Environmental Impacts

Luh Made Ratih Ratna Dewi,[1] Gusti Ayu Made Suartika,[1] Kadek Edi Saputra[1]

1 Master Program in Architecture, Development and Planning, Udayana University-Bali, Indonesia

Email: ratihluhmade@gmail.com

Abstract. Sidemen Distrct is one of the tourist attractions in Karangasem Regency. One of the prominent tourist attractions in Sidemen District from the past until now is the natural attractions of Iseh and Tebola as well as the sidemen songket weaving craft centre. Then, in 2010 began to expand tourism with the concept of "river tourism" namely tourism by making the Tukad Unda river as the main attraction. The development of tourism in Sidemen District is increasingly felt by the emergence of various tourism supporting facilities in the river border area. Of course, with the development of the construction of these facilities, it has an impact on the environment in the river border area, especially the spatial function seen from land use. The method used in this research is mapping analysis method with overlay technique. The results of the analysis explain that changes in land use that occur are clearly visible from the year of observation, namely in 2010 before the construction of tourism facilities until 2021. Changes in land use that occur lead to incompatibility with the direction of land use that has been determined. There are several land functions that have changed, including the function of plantation land/moor, and the function of agricultural land/rice fields. Of the two land functions, the one that shows the most changes because it is used for tourism facilities is the plantation/moor function of 5.43%. This percentage of land use change area is evidence that an activity such as tourism can have an influence on land function in an area. If this is left unchecked, it can result in arbitrary land use and not in accordance with the direction of the land function designation that has been stipulated in the spatial planning rules.

Keywords: riverfront development, environmental impact, tourism, uncontrolled land use

1. Introduction

River border areas tend to be in demand by investors to build tourism facilities and settlements. This tendency is caused by the attractiveness of the river border area as "river tourism" [1], which is an area that has a great opportunity in developing tourism potential whose attraction comes from the potential of the river itself [2]. The river border area is described as an area located on the outskirts of a dynamic village. From a spatial point of view, the river border area is a buffer zone between aquatic and terrestrial ecosystems [3]. However, the river border area also has challenges as a place or area with various social pressures and the dynamics of social change in it [4, 37, 38, 39].

The interest of investors in river border areas can encourage the phenomenon of changes in land function to the culture of the local community [5, 32, 33]. Changes in land use in an area can also be caused by several internal (local residents) and external factors. External factors can be in the form of industrial expansion, increasing ease of access to the location, and tourism development [6, 35, 36]. In planning studies, land use has two conflicting dimensions, namely the conservation aspect and the economic aspect. From the conservation aspect, ecological protection, water protection, and local area protection are needed.

From an economic point of view, land is very important as a factor of production, and land productivity is closely related to economic growth [7, 35, 36]. The occurrence of changes in land use patterns can expand the impact on employment, increase state income [8] creating a multiplier effect [9], and increase people's income [10,11]. In connection with the rapid development of tourism in Bali, many tourism activities use green land as tourism facilities [12]. Ecologically this can reduce the quantity of green open space that is actually intended for conservation functions. Tourist visits to Karangasem Regency each year have increased both foreign and domestic tourist visits. From 2015-2019 tourist visits to Karangasem Regency reached 2,991,709 people [13].

The conversion of land use for the construction of tourism facilities in the river border area mostly occurs in Sidemen District. Judging from the 2020 Karangasem Regency Spatial Plan, it states that the river border area has a function as a protected area. Protected area is an area designated with the main function to protect environmental sustainability which includes living natural resources and artificial natural resources, securing river flow, and preventing flood hazards. Seeing the strategic functions of the Tukad Unda river border area, it is necessary to take an action to control land use by looking at the trend of land development that occurs.

2. Land Use Change

Land is a natural resource that has a very broad function in meeting various human need. From an economic perspective, land is the main fixed input for various agricultural and non-agricultural commodity production activities. Developments and changes in land use patterns in residential and urban areas run and develop dynamically and naturally to nature [14], and are influenced by:

- Human factors, which consist of: the human need for a place to live, human potential, financial, socio-cultural and technological.
- City physical factors, including activity centres as urban growth canters and transportation networks as accessibility for ease of achievement.
- Landscape factors in the form of slope and land height.

3. Mapping Overlay Technique

Overlay is an important procedure in GIS (Geographical Information System) analysis. Overlay is an overlap between two maps that produces a new map unit of analysis and has the ability to place the graphics of one map on top of another map graphic and display the results on a computer screen or on a plot [15]. In short, overlay overlays a digital map on another digital map and its attributes and produces a combined map of the two that has attribute information from both maps [16].

4. Similar Research

So far, research on land use change has been carried out with a thematic, holistic, and integrative approach. However, in this study the researchers tried to use a spatial approach by using overlay mapping techniques to determine the magnitude of land use changes that occurred. In order to add insight and better understand the problems in this study, the researchers conducted a literature review of previous similar studies, namely the Changes in Spatial Functions As a Result of Tourism Development in Banjar Kedungu, Belalang Village, Kediri District, Tabanan Regency [17], Factors Affecting Functions Land in Subak Daksina, Tibubeneng Village, North Kuta District, Badung Regency [18], and Peri-Urban Spatial Transformation Around the Ciliwung River and Sustainability of River-Based Tourism Activities [19]. The first study describes changes in spatial functions seen in the context of tourism development and uses a spatial approach with mapping. Then, in the second study, we dissect the changes in land use that occur in subak land with the theory of land use changes. Meanwhile, the third study describes the pattern of changes in the area around the Ciliwung River as well as the evolution and direction of development of Saung Pustaka Air as alternative river-based tourism.

IOP Conf. Series: Earth and Environmental Science **903** (2021) 012013 doi:10.1088/1755-1315/903/1/012013

5. Method

This research was conducted in the Tukad Unda river border area in Sidemen District which covers Sidemen Village, Sangkan Gunung Village, Wisma Kerta Village. The research approach used is a qualitative approach and a quantitative approach. Limiting the problems that will be discussed in this study, the scope of this research is the conversion of land functions that occur due to the construction of tourism facilities in the Tukad Unda river border area, Sidemen District. The types of data in this study are qualitative and quantitative data. Data collection techniques used in this study include field observations and document review. Field observations to determine the use of existing land in the Tukad Unda river border area. The analytical method used in this research is to use the analysis of overlapping mapping (map overlay) with analytical instruments using GIS applications. The maps that will be analyzed are the land use maps in 2010 and the existing land use maps in 2021. The results of the overlay are then overlaid with the map of the land use plan that has been determined in the RTRW of Karangasem Regency. The types of data and data collection techniques needed can be seen in Table 1. as follows:

Table 1. Data Used

Data	Type	Source
RTRW Kabupaten Karangasem	Secondary Data	Publick Work and Spatial Planning Department Kabupaten Karangasem
Sidemen District Map	Secondary Data	Google Satelit
Sidemen Distric in Figure	Secondary Data	Statistic Kabupaten Karangasem
Land Use Map	Secondary Data	Kecamatan Sidemen
Existing Land Use Map	Primary Data	Survey, Google Satelit

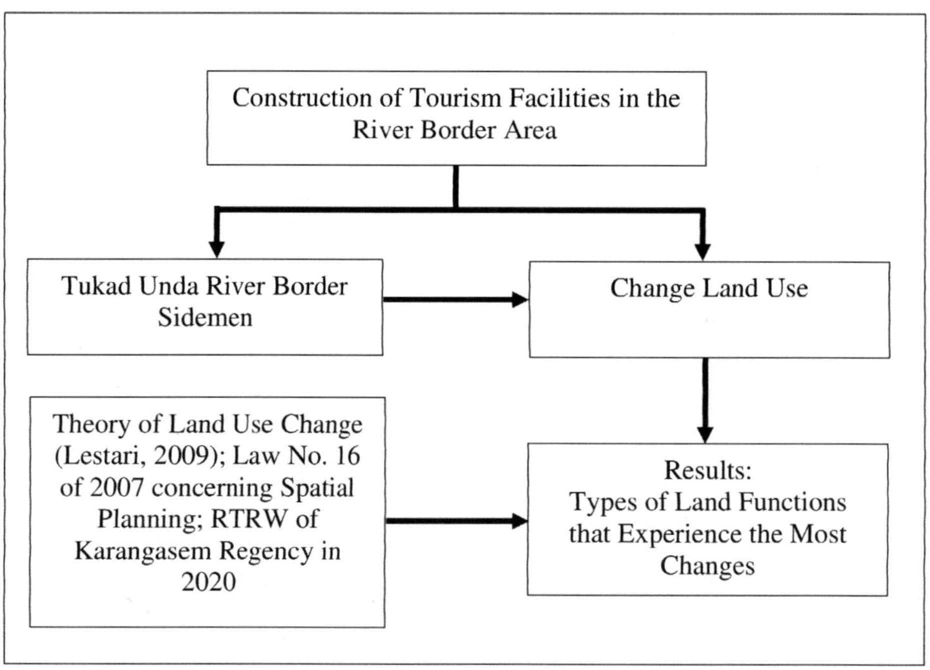

Figure 1. Research Model

The process of compiling the data obtained from the results of observations, interviews, literature studies, and documentation is described systematically and then the data analysis stage is carried out. In this study using descriptive analysis method. This research method is described through the research model scheme (Figure 1).

6. Tukad Unda River Border Area, Sidemen District

Geographically, the border area of Tukad Unda river is at coordinates: 8.484467°S and 115.444729°E in the administrative area of Sidemen District, Karangasem Regency, Bali Province. Physically, the research area of the Tukad Unda area has an area of about 63,684 hectares. The Tukad Unda area, Sidemen District, was chosen as the research location with the consideration that the river border area is a protected and cultivated area which is set at 50 (fifty) meters from the riverbed, while the development of tourism facilities such as hotels, villas and tourism facilities has been built in the river border area. the. To see more clearly the location of the Tukad Unda river border area, Sidemen District.

7. Result and Discussion

The impact of tourism in various areas in Bali is starting to threaten and cannot be avoided anymore. The incessant development of tourism brings negative influences and effects, especially for the natural and socio-cultural community [20,21]. Tourism activities that are closely related to social processes are slowly starting to affect all elements in tourism, including the community. The process of tourism activities in the Tukad Unda river border area, Sidemen District, raises hope for the community to get the maximum benefit from these activities. The hope to benefit from tourism activities is then realized by the community through the provision of tourism services to complement the needs of visiting tourists such as the construction of tourism facilities [22].

Tangibly, the growth of tourism facility development in the Tukad Unda river border area, such as homestays, villas, resorts, restaurants & SPAs, turned out to have a lot of negative impacts. And if it is associated with the manifestation and latency approach, this phenomenon is latency or an unexpected form of the tourism development process that naturally always accompanies or appears [23,24,25]. The negative impact is the impact on the land use of plantations/moor and rice fields whose existence is threatened because the development of community-managed tourism facilities is built on a river border area with land functions as plantations/moorlands and rice fields. To be further elaborated in the following explanation:

7.1 Existing Land in the Tukad Unda River Border Area, Sidemen District

Land use in the Tukad Unda river border area, Sidemen District, is dominated by land use for food crops (gardens/moor) and wetland areas (rice fields). Based on the results of the inventory of existing land use in the Tukad Unda river border area, it can be seen that the land use conditions on the Tukad Unda river border are not only developed for food crop activities (gardens/moorlands) and wetland agricultural cultivation (rice fields), but other types of uses have developed. such as areas for tourism facilities, settlements, and MSMEs. Land use in the Tukad Unda river border area in the existing year can be seen in Table 2.

The research area of the Tukad Unda river border area has an area of ± 63,684 covering the administrative areas of Sidemen Village, Sangkan Gunung Village, and Wisma Kerta Village. From the data in the table above, it can be seen that the dominant land use in the Tukad Unda river border area in 2021 is the use of plantation/moor land with an area of 38.190 Ha or 59.67% of the total area of the study area. Land use in the Tukad Unda river border area, Sidemen District, can be seen through the use map in Figure 2.

Table 2. Land Use in the Tukad Unda River Border Area, Sidemen District in 2021 (Existing)

Type of Land Use	Land Area in 2021 (Ha)
Plantation/Moor	38.190
River	12.039
Rice Field	8.055
Settlement	0.637
Tourism	4.042
MSMEs	0.718
Amount	**63.684**

Source: [26]

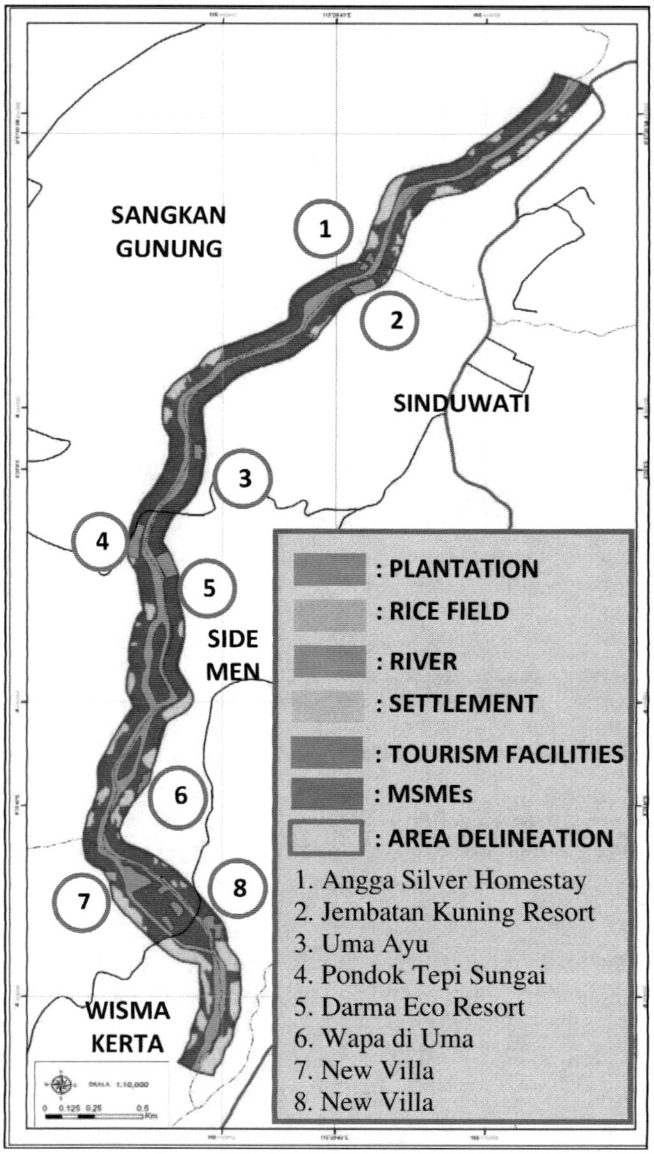

Source: [27]

Figure 2. Land Use in the Tukad Unda River Border Area, Sidemen District in 2021

IOP Conf. Series: Earth and Environmental Science **903** (2021) 012013 doi:10.1088/1755-1315/903/1/012013

7.2 Land Function in the Tukad Unda River Border Area, Sidemen Distric

The development of tourism in Sidemen District involves a wide range of issues, ranging from experience during the trip to the condition of the supporting facilities. The quality of a tourist area depends on the attractions, transportation, accommodation, places to eat, and entertainment of tourists who are far from where they live. To accommodate all these activities, it is necessary to build various physical facilities. This means that tourism causes changes to land use and function [28]. In its development, land use with a function as carrying capacity and the environment in the Tukad Unda river border area cannot maintain the composition of its area because of the factors that cause changes. One of them is because of the construction of tourism facilities. Changes that occur more clearly can be seen from the land use in the Tukad Unda river border area in the 11year period before the start of the construction of tourism facilities. In this study, the year of observation taken was 2010. In 2010 it was clear that land use in the Tukad Unda river border area was still dominated by land use with the function of environmental carrying capacity, which can be seen in its distribution in Table 3. below:

Table 3. Land Use in the Tukad Unda River Border Area, Sidemen District, 2010

Type of Land Use	Land Area in 2020 (Ha)
Plantation/Moor	42,305
River	12,039
Rice Field	9,170
Settlement	0,167
Tourism	0
MSMEs	0
Amount	**63,684**

Source: [26]

Based on table 3. above, it is known that land use in the Tukad Unda river border area in 2010 was dominated by plantation/moor land use, which amounted to 42,305 Ha or 66.42% of the total area of the study area. In 2010 there has not been any development of tourism facilities. This indicates that before the development of tourism, the main occupation of the people in the Tukad Unda river border area, Sidemen District, was as farmers and sidemen songket weaving craftsmen.

If you compare the land use map before the construction of tourism facilities (in 2010) with the land use map after the construction of tourism facilities (in 2021), it can be clearly seen that there are differences in land use which can be illustrated in Table 4. and Figure 3. as follows:

Table 4. Comparison of Land Use in 2010 and 2021 in the Tukad Unda River Border Area, Sidemen District

Type of Land Use	Land Area in 2010 (Ha)	Land Area in 2021 (Ha)	Percentage (%)	Info
Plantation/Moor	42,305	38,190	5,43%	Down
River	12,039	12,039	0%	Same
Rice Field	9,170	8,055	1,35%	Down
Settlement	0,167	0,637	0,01%	Up
Tourism	0	4,042	6,74%	Up
MSMEs	0	0,718	0,03%	Up
Amount	**63,684**	**63,684**		

Source: [26]

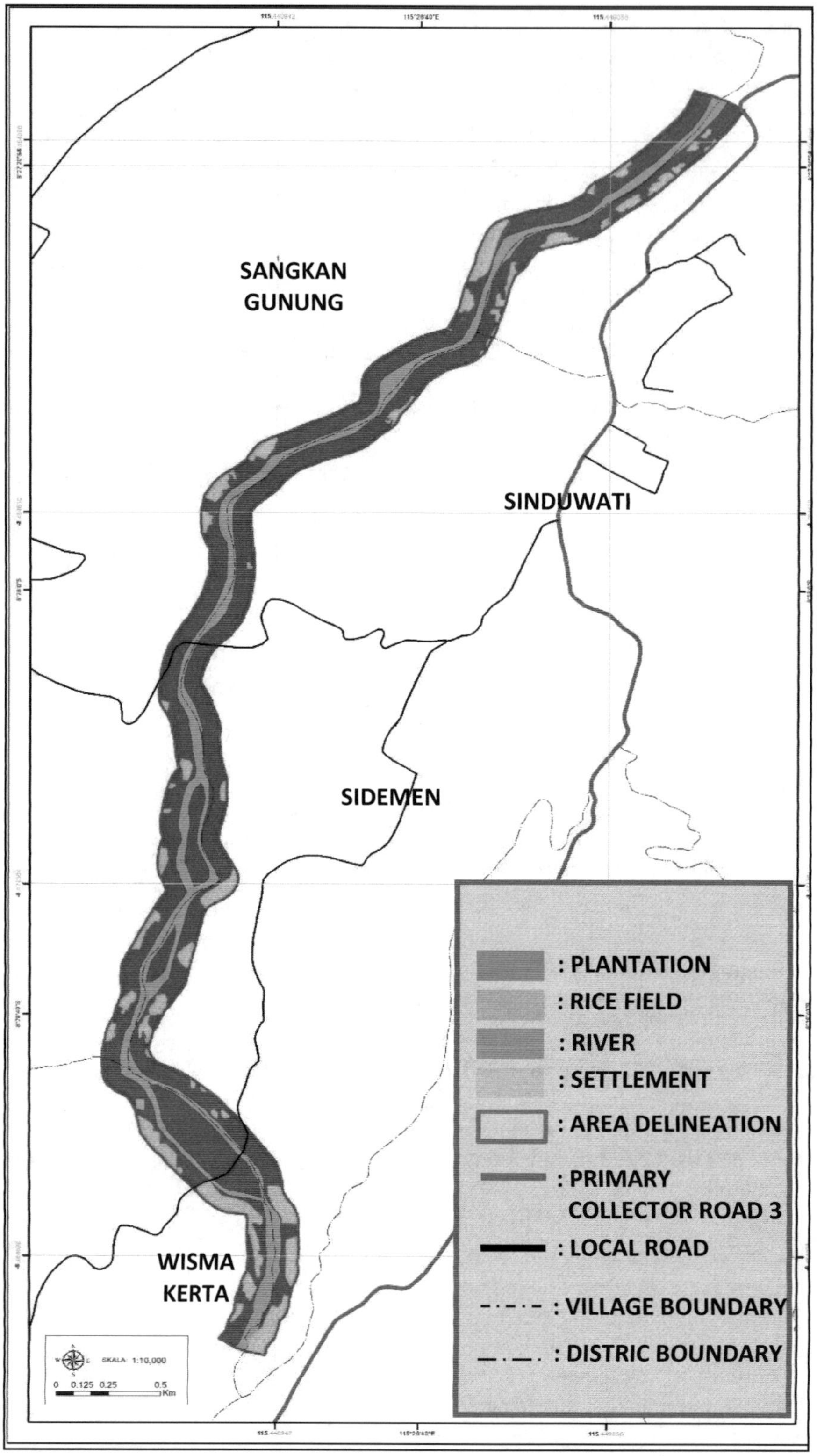

Source: [29]

Figure 3. Land Use in the Tukad Unda River Border Area, Sidemen District in 2010

IOP Conf. Series: Earth and Environmental Science **903** (2021) 012013 doi:10.1088/1755-1315/903/1/012013

Table 4. shows that over a period of 11 (eleven) years (2010-2020) the Tukad Unda river border area, Sidemen District has experienced very rapid land use changes. For more details can be described in Figure 4.

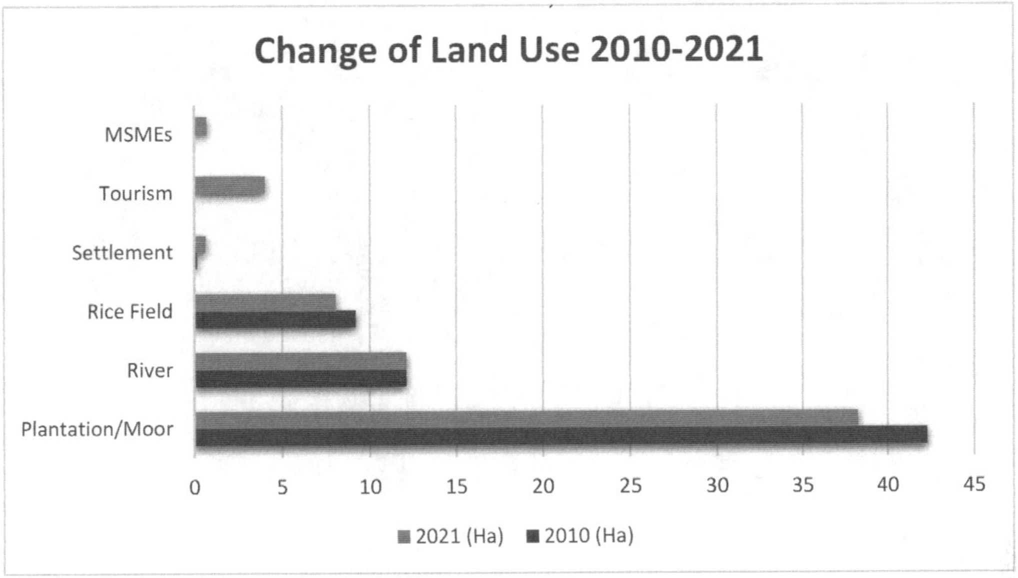

Source: Autho [30]

Figure 4. Changes in Land Use in 2010-2021

From Table 4. and Figure 5. above, it can be analyzed that there are changes in several land functions in the Tukad Unda river border area which is the impact of the construction of tourism facilities. The construction of this tourism facility utilizes plantation/moor land which is a food crop area with a protected function. This condition is further supported by the data analysis that has been carried out regarding land use changes that occurred in the Tukad Unda river border area before and after the construction of tourism facilities. In accordance with the table above, it can be seen that the plantation area/moor area has decreased in area, from 42,305 Ha in 2010 to 38,190 Ha in 2021.

This shows that there is land use in the plantation area which causes the area to decrease. Not only the area of plantations/moor which has decreased in area, the same case also occurs in the use of agricultural land/rice fields. Agricultural land/rice fields decreased from an area of 9,170 Ha in 2010 to 8,055 Ha in 2021. Then, the use of land for the construction of tourism facilities showed an increasing number from an area of 0 Ha in 2010 and until now in 2021 it has grown to 4,042 Ha. This proves that the development of tourism facilities in the Tukad Unda river border area, Sidemen District has increased. Tourism facilities in this river border area can be in the form of homestays, villas, resorts, restaurants & SPAs, and food stalls. Spatially, the areas undergoing changes can be compared through Figure 5. as follows:

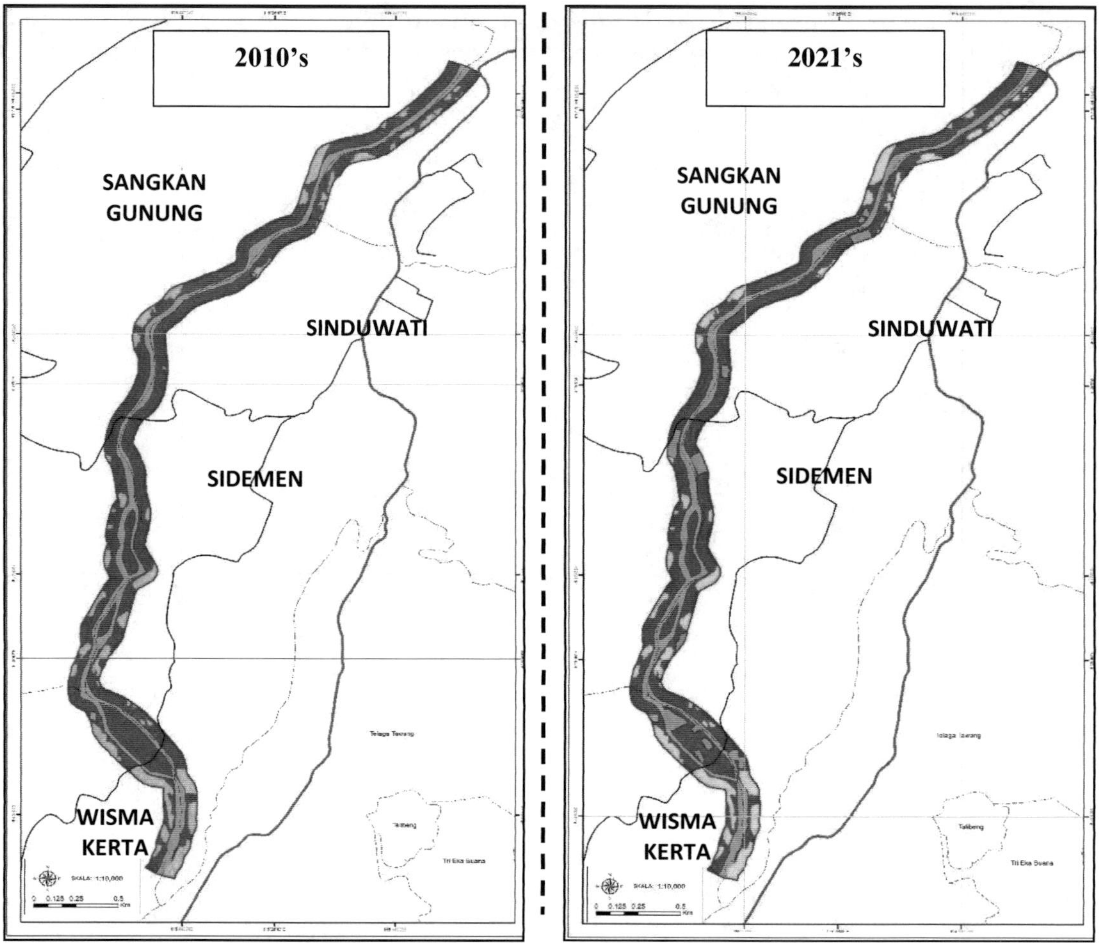

Source: [26]

Figure 5. Comparison of Land Use in 2010 and 2021

Figure 5. above shows a comparison of land use functions in the 2010-2021 period. This can be seen in the garden/moor land which decreased by 5.43% and the paddy field decreased by 1.35%. Meanwhile, for land that has increased its use, tourism is 6.74% and MSMEs is 0.03%. From these figures, if accumulated, it shows that the development of tourism facility development in the Tukad Unda river border area, Sidemen District, in addition to having an impact on the community's economy, also has an impact on land use with a tendency to decrease open land that functions as an ecological function. When viewed from a spatial perspective, it is analyzed using the Arc GIS 10.5 application juxtaposed with satellite imagery in 2021 and then overlaid with the spatial pattern of the Karangasem Regency RTRW. At this stage of analysis, the existing land use, area, and percentage of spatial use will be determined according to and not in accordance with the spatial plan (RTR). The area of the river border is measured as far as 50 m from the left and right of the river without embankments for rivers outside the residential area, using GIS, the research area is ± 63,684 ha.

The results of the overlay analysis above obtained 3 (three) types of activities that utilize space in the Tukad Unda river border area, Sidemen District which are not in accordance with their designation. Utilization of space on the river border that is not in accordance with the spatial plan includes (homestays, villas, resorts, restaurants & SPAs, and food stalls), settlements, and MSMEs located within the radius of the river border with an area of 5,397 ha or 6.78% of the total inappropriate utilization. the total area of the research area as presented in Table 5. and Figure 6.

Table 5. Incompatibility of Land Use in the River Border Area Tukad Unda, Sidemen District

No	Utilization Type	Area	
		Ha	%
1	Tourism Facility	4,042	6,74%
2	Settlement	0,637	0,01%
3	MSMEs	0,718	0,03%
	Amount	**5,397**	**6,78%**

Source: [26]

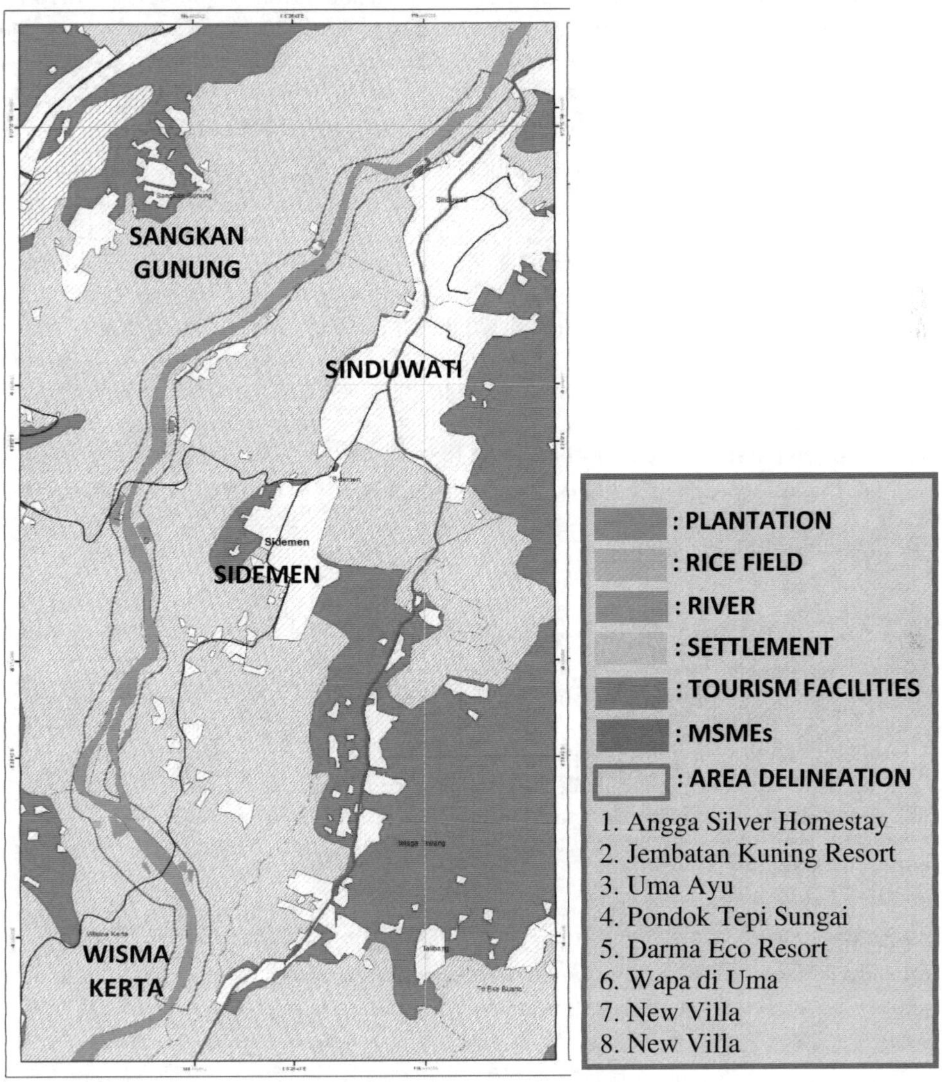

Source: [31]

Figure 6. Comparison of Land Use in 2010 and 2021

The discrepancy in the use of space in the Tukad Unda river border area, Sidemen sub-district, from the results of questionnaires and interviews, can be concluded due to several things including:

IOP Conf. Series: Earth and Environmental Science **903** (2021) 012013 doi:10.1088/1755-1315/903/1/012013

- Lack of public understanding of the function of river borders
- Conflicts between activities/sectors in the use of river border space
- There is no detailed spatial plan (RDTR) that is legalized
- Weak enforcement of the Karangasem Regency RTRW Regional Regulation

8. Conclusions and Recommendations

This study This study shows that the development of tourism facilities in the Tukad Unda river border area, Sidemen District, can arise through individual or group interest in the issues and challenges of the conditions that develop around them. Although, at the same time, this also encourages the emergence of the phenomenon of land function changes/shifts that occur due to the need to develop a tourism facility that can support tourism activities in Sidemen District. The land function that experienced the most changes in function was the function of plantations/moorlands from 42,305 Ha in 2010 to 38,190 Ha in 2021 or decreased by 5.43%. The change in land use function in the Tukad Unda river border area from a protected area function to a built area appears to be very significant in the downstream part of the river compared to the other side of the river. The related parties who carry out this activity have their own reasons for using land that is actually not in accordance with the designation plan.

The suggestions that can be conveyed in this research are: a) The need to maintain and maintain the environment in the river border area as an object and tourist attraction in order to continue to provide a beautiful and original atmosphere while still providing value for benefits for the community, b) Preserving the attractiveness of the border area river by compiling in detail an integrated tourist attraction spatial plan, c) Researching alternative regional developments, especially western to reduce tourism spatial agglomeration in the Tukad Unda river border area so as to reduce the environmental burden, d) Conduct training and socialization to stakeholders related to tourism development so that In the future, Sustainable Tourism Development will be created. Tourism that has a positive impact on social, economic, and cultural aspects without destroying the environment, and e) This research is expected to enrich the reference literature in the field of regional development, especially rural and sub-district areas regarding land use plans and controls that must be carried out, and the results of this research is expected to be a useful reference for understanding and further developing this kind of research or a more in-depth study.

9. References

[1] Faiz M 2017 *Potensi Daerah Aliran Sungai Ci Manuk Bagian Hilir Sebagai Objek Wisata di Kecamatan Indramayu dan Kecamatan Sindang* (Bandung: Pascasarjana Universitas Pendidikan Indonesia)

[2] Fandeli C 1995 *Dasar-Dasar Manajemen Kepariwisataan Alam* (Yogyakarta: Liberty)

[3] Anonym 2020 *Peraturan Kementerian Pekerjaan Umum dan Perumahan Rakyat Nomor 28/PRT /M/2015* (Jakarta: Menteri Pekerjaan Umum dan Perumahan Rakyat, 1-4)

[4] Iaquinta D 2000 *Defining Periurban: Understanding Rural-Urban Linkages and Their Connection to Instutional Contexts* (Working Paper)

[5] Pratiwi W.D 2019 *Transformasi Fisik, Teritorial, dan Budaya dalam Permukiman* (Bandung: ITB Press)

[6] Pratiwi W.D 2019 *Sustainable Settlement Development: Land Use Change in Lakeside Tourism of Bandung in Equity, Equality, And Justice in Urban Housing Development* (KnE: Social Sciences, 883-897)

[7] Gun C.A 1988 *Tourism Planning (2nd Ed.)* (New York: Taylor & Francis)

[8] Schober B 2010 *Assessing land use change impacts – a comparison of the SENSOR land use function approach with other frameworks* (New York: Taylor & Francis, 159-160)

[9] Walker J.R 2011 *Tourism Concept and Practice* (New Jersey: Prentice Hall)

[10] Goeldner, C. R. 2003 *Tourism: Principles, Practices, and Philosophies (9th Ed.)* (New *Jersey*: John Wiley & Sons, Inc)

[11] Henderson J 2004 *Widening Accomodation Choice: The Potential of Homestay in Singapore* (Bandung: ASEAN Journal on Hospitality and Tourism,3, 1-9)
[12] Suwena Ketut 2016 *Pengembangan Pariwisata Hijau Sebagai Upaya Pemberdayaan Masyarakat di Desa Pemuteran Kabupaten Buleleng* (Denpasar: Sanastek, 1-2)
[13] Anonym 2020 *Karangasem Dalam Angka 2019* (Karangasem: Badan Pusat Statistik Kabupaten Karangasem, 333-340)
[14] Lestari T 2009 *Dampak Konservasi Lahan Pertanian Bagi Taraf Hidup Petani* (Institut Pertanian Bogor: Jurnal Agribisnis Perdesaan, 95-107)
[15] Aronoff 1989 *Geographic Information System* (Canada: A Management Perspective WDL Publication Ottawa)
[16] Hasyim W 2011 *Urban Land Use Change Analysis Using Temporal Multispectral Imagery and Image Diffrence* (Serialonline: https://awhasyim.wordpress.com/category/remote-sensing/)
[17] Rapiana E 2017 *Perubahan Fungsi Spasial Akibat Perkembangan Pariwisata di Banjar Kedungu, Desa Bebalang Kecamatan Kediri Kabupaten Tabanan* (Denpasar: Pascasarjana Universitas Udayana)
[18] Suputra A 2012 *Faktor-Faktor yang Mempengaruhi Alih Fungsi Lahan Studi Kasus di Subak Daksina, Desa Tibubeneng, Kecamatan Kuta Utara, Kabupaten Badung* (Denpasar: Journal of Agribusiness and Agrotourism, 1, 61-68)
[19] Sesotyaningtyas M 2018 *Transformasi Spasial Peri-Urban di Sekitar Sungai Ciliwung dan Keberlanjutan Kegiatan Wisata Berbasis Sungai* (Denpasar: Jurnal Space, 8, 17-26)
[20] Soerjono S 2005 *Sosiologi Suatu Pengantar* (Jakarta: Raja Grafindo Persada)
[21] Pitana Gayatri 2005 *Sosiologi Pariwisata* (Yogyakarta: Andi)
[22] Faizun M 2009 *Dampak Perkembangan Kawasan Wisata Pantai Kartini Yarakat Setempat di Kabupaten Jepara* (Semarang: Pascasarjana Universitas Diponogoro)
[23] Goodall B 1996 *Aspect of Land Economics* (Bath: Ptmann Press)
[24] Saladin D 2003 *Intisari Pemasaran dan Unsur-Unsur Pemasaran Cetakan Ketiga* (Bandung: Linda Karya)
[25] Dipayana A 2015 *Dampak Pariwisata Terhadap Alih Fungsi Lahan di Desa Tibubeneng Kecamatan Kuta Utara Kabupaten Badung* (Denpasar: Jurnal Destinasi Pariwisata, 3, 58-66)
[26] Anonym 2010;2021 *Peta Rupa Bumi Indonesia (RBI)* (Bogor: Badan Informasi dan Geospasial)
[27] Anonym 2021 *Peta Rupa Bumi Indonesia (RBI)* (Bogor: Badan Informasi dan Geospasial)
[28] Zaenuri M 2012 *Perencanaan Strategis Kepariwisataan Daerah* (Jakarta: e-GOV *Publishing, 11-115)*
[29] Anonym 2010 *Peta Rupa Bumi Indonesia (RBI)* (Bogor: Badan Informasi dan Geospasial)
[30] Ratih Ratna 2021 *Dampak Pembangunan Fasilitas Pariwisata Terhadap Lingkungan Fisik di Kawasan Sempadan Sungai Tukad Unda Kecamatan Sidemen Kabupaten Karangasem, Bali* (Denpasar: Pascasarjana Universitas Udayana)
[31] Anonym 2020 *Rencana Tata Ruang Wilayah Kabupaten Karangasem Tahun 2020* (Karangasem: Bupati Karangasem)
[32] Suartika, G A M 2013 *Open Space Traditions, Development and Modernity: The Case of Bali* Journal Nasional Terakreditasi Mudra: Journal Seni Budaya Vol. 28, No. 2 December 2013, (Institut, Seni Indonesia Denpasar, UPT Penerbit, p: 283-299)
[33] Suartika, G A M 2007 *Perencanaan keruangan, wujud komunikasi kepentingan dalam pembangunan Journal Natah, Vol 5, No. 2* (Denpasar, Bali: Laboratorium Permukiman, Jurusan Arsitektur, Fakultas Teknik, Universitas Udayana, p: 71-78)
[34] Suartika, G A M. Cuthbert, A R, Putra, G M, Saputra, K E 2020 *Public Domain and Cultural Legacy: The Governance of a Sacred and Vernacular Cultural Landscape in Bali* (ISVS e-Journal Vol 7, Issue 2, p:1-22)
[35] Suartika, G A M, Zerby, J, Cuthbert, A R 2018 *Doors of Perception to Space – Time - Meaning: Ideology, Religion, and Aesthetics in Balinese Development' Journal Space and Culture, Vol 21 (4) November, p: 340-357* (Sage Publisher)

[36] Cuthbert, A R, Suartika, GAM 2014 *The Theatre of the Universe-Culture and Urban Design* Journal of Urban Design, Vol. 19 No. 3, 2014, p: 275-277 (Routledge, Taylor and Francis Group, United Kingdom)

[37] Suartika, G A M, Permanasuri, N P D A and Saputra, K E 2019 *Provision of A Sustainable Public Space: Lebih Coastal Area in the Aftermath of Rampant Abrasions*, in (IOP Conf. Ser.: Earth Environ. Sci. 396 012029, p:1-12)

[38] Suartika, GAM, Ledysianty, I A S O and Saputra, K E 2019 *Land Use Change Along Lebih Coastal Line, Entailing Its Designation as A Zone for Tourist Development* in (IOP Conf. Ser.: Earth Environ. Sci. 396 012014, p: 1-16)

[39] Suartika, G A M 2015 *Sand, Sea and Ceremony: Conflict over the Littoral Public Realm in Sanur, Bali* (Procedia - Social and Behavioural Sciences 179 (2015) p: 128-140, Elsevier)

10. Acknowledgement

Authors are grateful to the Indonesian Ministerial for Education, Culture, Research and Technology for enabling the publication of this article by providing research grants (with a contract No of: B/136-10/UN14.4.A/PT.01.05/2021, dated 29 June 2021).

IOP Conf. Series: Earth and Environmental Science **903** (2021) 012014 doi:10.1088/1755-1315/903/1/012014

Carrying Capacity of Land in Bali based on Ecological Footprint

A A B Bayu Baskhara Hutama[1], A K Muktiwibowo[1], Widiastuti[1] and K E Saputra[1]

[1] Masters Program in Architecture, Development and Planning

Email: gungbaskhara@gmail.com

Abstract. In 2000-2010, based on data from the Population Census in Bali, the average population growth rate was 2.15% per year, this value was greater than the average population growth rate at the national level of 1.49% per year. The increasing number of population growth results in increased development. This causes an increase in the need for land to accommodate these development activities. The reduced availability of productive land because it is used for development facilities has resulted in reduced land carrying capacity in Bali. The purpose of this study was to analyze the value of the carrying capacity of land in Bali using the Ecological Footprint analysis approach. The research methodology is quantitative focusing on secondary data analysis, based on the mathematical calculation method developed by the Global Footprint Network, which has the term biocapacity which means supply and ecological footprint which means demand. After comparing biocapacity and ecological footprint, the results show that the carrying capacity of land in Bali is generally in deficit. Things that can be done to increase the carrying capacity of land are to reduce the rate of population growth, increase land productivity by intensifying agricultural land, and be wiser in carrying out land conversion in the future.

Keywords: land carrying capacity, ecological footprint, land biocapacity and productivity

1. Introduction

This article focuses on the study of environmental carrying capacity or can be referred to as carrying capacity, especially on the aspect of land carrying capacity in Bali using an ecological footprint analysis approach. Similar studies have been carried out in other areas as the object of study, such as Pattalassang District, Takalar Regency [1] and Gresik Regency [2]. However, similar research on the carrying capacity of land has never been carried out in Bali as the object of study. Another thing that supports this research is the average annual population growth of Bali during the last decade of 2.15%, which exceeds the average rate of growth of Indonesia's population per year in the same period of 1.49% (BPS Province of Bali., 2020). This population growth has an impact on increasing population needs such as housing, work places and consumer goods. This triggers an increase in activities in all aspects of community life, both physical, economic, social, political and cultural development, which leads to an increase in the need for land as a place for all these activities and a decrease in the carrying capacity of land in Bali.

According to Rees, the carrying capacity of the environment and environmental conditions are strongly influenced by changes in land use. In the context of ecology, environmental carrying capacity is defined as the ability of an ecosystem or environment and the resources and services in it to ensure the survival of a number of populations or communities in it [3]. In Article 1 of the Minister of Environment Regulation number 17 of 2009, it is stated that the carrying capacity of the environment is the ability of the environment to support the life of living things, both humans and other living things.

One part of the carrying capacity of the environment is the carrying capacity of the land, which can be interpreted as the capacity of the land to accommodate humans who live on it and all these human activities. The carrying capacity of this land is also one of the basic principles to achieve sustainable development.

In order to support the development of the island of Bali as a province that puts forward the principles of sustainable development, it is very important to know the status of the carrying capacity of land in the province of Bali. By understanding the status of the carrying capacity of this land, preventive and strategic efforts can be designed early so that Bali can be maintained as an ecosystem that is able to support a harmonious life in accordance with the Tri Hita Karana concept.

In order to measure the carrying capacity of land, the approach used in this research is the principles of ecological footprint analysis. This method is used as an approach in measuring the boundaries of sustainable ecosystems, through measuring population demand for the environment or nature expressed in units of global biocapacity area. The ecological footprint approach, a concept first expressed by William Rees and Mathis Wackernagel in 1996, according to Febrianto has now become a very important reference in measuring global sustainability [4].

This article consists of several parts, namely, an introduction that discusses and demonstrates the background of the research, a research methodology that contains the research approach used, research results and a discussion that contains data and results from the research, the last is a closing which contains conclusions from the results. study. Through this article, it is hoped that the public can find out the status of the carrying capacity of land in Bali in 2019, and the importance of the carrying capacity of land on the environment in Bali, so that the community can take steps that can support the value of the carrying capacity of land so as to create sustainable development in Bali.

2. Literature review

2.1. Sustainable development

Sustainable development has been widely accepted by many countries in the world as a human target and a new model of development in their country [5]. In simple terms, sustainable development according to Ms. Brundtland aims to meet the needs of humans living today without compromising the ability of future generations to meet their needs for life and development [5][6]. This concept will involve at least four key elements, namely population, resources, environment and economy, known as PREE. Most of the existing studies believe that economic development should be based on good ecological environmental cycles [7]. These elements are very complex sub-systems so that the interaction between these sub-systems in the sustainable development system causes the sustainable development model to become a very complex system [8], both in understanding, assessing and choosing strategies to achieve it [9] . In 2030, there are 17 sustainable development goals that have been set as a world transformation agenda to address the challenges of various interests in the sustainable development sub-system, namely no poverty, no hunger, health and well-being, quality of education, quality of gender, clean water and sanitation, affordable clean energy, decent work and economic growth, industry-innovation and infrastructure, reducing diversity, sustainable communities and cities, responsible consumption and production, action on climate, underwater life, life on land, peace- justice and strong institutions, and partnerships for achieving goals [10]. Taking into account the many targets of sustainable development and the existence of conflicts in implementing them, assessing the success of implementing sustainable development is very complex and there must be trade-offs between sub-systems and between targets [5][11]. The use of the concept of carrying capacity (CC) is not accurate enough to assess sustainable development [5]. Carrying capacity is simply defined as the maximum population that an ecosystem can support. Carying capacity from the point of view of regional or urban planning, is a tool to measure the achievement of the implementation of the concept of sustainable development, because CC can be used to determine the level of human activity, population growth, land use patterns and developments, physical development that can be supported by the urban environment without causing degradation. serious environment [12]. Various methods are used to estimate the carrying capacity of an area that is non-linear. Principle component analysis (PCA) is used in research conducted by Liu et al [13] to estimate the carrying capacity of an element in the concept of sustainable development in various regions in China. The results of his research indicate that the carrying capacity of water resources and environmental quality is relatively low compared to the carrying capacity of infrastructure which is the highest among the components considered. Furthermore, Wei et al [14] used urban-carrying capacity as a benchmark for sustainable urban development in Beijing, while Andreson et al [15] used carrying capacity for modeling population development. In Indonesia, carrying capacity is used to assess the sustainability of an inland fishery area based on the productivity of chlorophyll and phosphate in the sub-system [16]. In addition, the ecological footprint is also widely used to assess the sustainability status of an area or its elements such as the carrying capacity of agricultural land on the island of Java [17], the environmental carrying capacity (DDL) of Tamangapa Village [4], the carrying capacity of the city of Bandung [18], optimization land use in Gresik [2], land carrying capacity status in Pattalassang sub-district [1], to determine the status of urban food consumption [19]. On the other hand, Wijaya et al [20] used satellite data analysis to assess land capacity in Kotabaru, South Kalimantan.

Regional planning regulations in Indonesia take into account environmental and spatial/regional issues. The concept of carrying capacity which includes supportive carrying capacity (SCC) and assimilative carrying capacity (ACC) has been clearly defined in it [8]. SCC is the capacity/ability of the environment to support human life and other living things in it. Meanwhile, ACC is the capacity/ability of the environment to absorb elements, energy and/or components that enter the environment, either independently or with human intervention. The goal of spatial planning in Indonesia is the effective and efficient use of space, which is very important in sustainable environmental management, prevention of space wastage and degradation of space quality. Spatial planning is based on the characteristics of the available land, the carrying capacity of the land and the relevant supporting technology. Spatial planning in Indonesia is expected to result in the suitability of space utilization, harmony, and balance of the sub-systems in it [8].

2.2. Biocapacity and ecological footprint
The supply aspect describes the ability of an ecosystem or environment to support the life of living things which is called biocapacity. The ecological footprint is an analytical tool to measure and communicate the impact of resource use on the environment, the ecological footprint is one of the coefficients needed to analyze the carrying capacity of the environment based on the ecological footprint, the ecological footprint also describes the demand aspect. To determine the value of biocapacity, several variables are needed, namely land area of each category of bioproductive land use, equivalence factor, and yield factor (harvest factor). Determining the value of the ecological footprint also requires several variables, namely the amount of product production per each land category, the number of exports and imports of food products from each land category, product productivity, equivalence factor, and yield factor (harvest factor).

2.3. Land carrying capacity in ecological footprint
Based on the publication of the Living Planet Report [21], the comparison between biocapacity (supply) and ecological footprint (demand) can reflect the carrying capacity of an area's land. In the calculation, if the ecological footprint is greater than the biocapacity, there will be an overshoot, which means that the carrying capacity of the land has been exceeded. In this condition, there is an ecological deficit or unsustainable status. Conversely, if the ecological footprint is smaller, then there is a certain amount of biocapacity in nature that is reserved to support future life (ecological debt) or has a sustainable status. The value of the carrying capacity of the land can be analyzed from the results of the value of the analysis of biocapacity divided by the value of the results of the analysis of the ecological footprint.

3. Method
The research methodology used is quantitative research that focuses on secondary data analysis, where secondary data is valid data relating to the variables to be analyzed. The secondary data needed is from the Central Statistics Agency which contains the amount of production and product productivity from each category of bioproductive land in Bali in 2019, then data from the National Land Agency for the use of each land category in Bali in 2019, and data from the National Land Agency for the use of each

land category in Bali in 2019. from the Bali Province Trade and Industry Office to find out about exports and imports of food products in Bali in 2019.
The data analysis used in this study is based on a mathematical calculation method called Ecological Footprint Analysis. This method has been developed by the Global Footprint Network (GFN) (2019) in the Working Guidebook to the National Footprint and Biocapacity Accounts [22].

3.1. Biocapacity
Biocapacity for all land categories was calculated using the following equation:

$$BC = A \times YF \times EQF \tag{1}$$

BC = Biocapacity
A = Land use area per land category
YF = Yield factor
EQF = Equivalence factor

3.2. Ecological footprint
The ecological footprint can be calculated using the following equation:

$$EF = \frac{P}{Y_N} \times YF \times EQF \tag{2}$$

Information:
EF = Ecological footprint
P = Total consumption of the product (after adding imports and deducting exports) (tons/year), if not available, land area can be used
YN = Productivity of the calculated area (tons/ha/year)
YF = Yield factor
EQF = Equivalence factor

3.3. Equivalence factor
The value of the equivalence factor in this study refers to the data in the Working Guidebook to the National Footprint and Biocapacity Accounts [22] as presented in Table 1.

Table 1. Equivalence Factor.

No	Land Use	EQF
1	Agricultural land	2,52
2	Forest	1,29
3	Meadow/Farm/Field	0,46
4	Aquatic Land	0,37
5	Built-up Land	2,52

3.4. Yield Factor
Based on the method developed by the Global Footprint Network, the harvest factor or yield factor can be calculated by dividing the productivity of a land category in the calculation area with the average productivity of the same land in the world in the same year, the harvest factor can be calculated by the following formula.

$$YFL = \frac{YNL}{YWL} \tag{3}$$

Information:
YFL = *Yield factor (harvest factor) for land category L*
YNL = Land productivity in the calculation area for land category L (tons/ha/year)
YWL = World land productivity for land category L (tons/ha/year)
There are several assumptions for calculating the yield factor in completing equation (3) above, the assumption in question is that the built-up land has the same yield factor as agricultural land, due to the

assumption that urban land usually converts agricultural land. Fishery land has a yield factor value of 1 or the productivity of the world's fishery land is the same as the calculation area, this is assumed due to the limitations of world fishery data, especially in remote countries or regions [2].

3.5. Land Carrying Capacity

To find out the value of environmental carrying capacity based on the ecological footprint in Bali Province, it can be calculated by the following formula.

$$DDL = BC \,/\, EF \tag{4}$$

Information:
DDL = Land carrying capacity
BC = Biocapacity
EF = Ecological footprint
Based on equation (4) above, if the carrying capacity of the land (DDL) > 1, it can be interpreted as a surplus carrying capacity condition, where the ecosystem is able to support human life living in it. If DDL < 1, it can be interpreted as an overshoot or deficit carrying capacity condition, where the land is not able to support the human life that lives in it [18].

3.6. Regional Development Potential

Based on the DDL value, it can be determined the optimal number of residents, the population is not accommodated, the optimal land area, and additional land area, described in Table 2. What is meant by optimal is the amount of biocapacity equal to the ecological footprint. Thus, the optimal population means the maximum population that can be supported by the ecosystem, while the optimal land area means the minimum land area needed to support the population living on it [4].

Table 2. Regional Development Potential.

Formula	Information
$JPO = DDL \times JP$	JPO : Optimal population size that can be supported JP : Population size
$JPTT = (1 - DDL) \times JP$	JPTT : The number of people who cannot be supported JP : Population size
$LLO = L_{total} \times DDL^{-1}$	LLO : Optimal land area
$LLT = \left(\frac{1}{1 - DDL}\right) \times L_{total}$	LLT : Additional land area needed to support population

4. Data, discussion, and results/findings

4.1. Land Use in Bali Province in 2019

Bali has an area of 563,666 hectares with a population of 4,336,900 people in 2019. The following is data on land use in Bali based on data from the State Land Agency of Bali Province in 2020 [23], presented in Table 3.

Table 3. Land Use in Bali Province in 2019

No	Land Use	Area (Ha)
1	Agricultural land	89.138,42
2	Forest	106.241,34
3	Tegal/Field	144.363,10
4	Plantation	137.440,64
5	Inland Water	5.553,53
6	Built-up Land	80.928,78
	Total	563.666

IOP Conf. Series: Earth and Environmental Science **903** (2021) 012014 doi:10.1088/1755-1315/903/1/012014

4.2. Product Production and Productivity in Bali in 2019

The following is data on product consumption per land use category in Bali. The data used are production and product productivity data in 2019 from the Bali Provincial Central Statistics Agency [24], and food export and import data in 2019 from the Bali Provincial Industry and Trade Office [25] which are presented in Table 4.

Table 4. Product Production and Productivity in Bali Province in 2019.

No	Land Use	P (ton)	YN (ton/ha/Year)
1	Agriculture	714.837,3	6,74
2	Tegal/Field	727.536,87	5,09
3	Plantation	94.853,15	4,47
4	Inland Water	15.538	2,80

4.3. Yield Factor Analysis

The following analysis of the yield factor, using equation (3), is presented in Table 5.

Table 5. Yield Factor Analysis

No	Land Use	YNL	YWL	YFL
1	Agriculture	6,74	4,63	1,46
2	Forest	-	1,82	-
3	Tegal/Field	5,09	6,19	0,82
4	Plantation	4,47	6,19	0,10
5	Inland water	2,80	2,80	1
6	Built-up Land	-	-	1,46

4.4. Biocapacity Analysis

The following analysis of biocapacity, using equation (1), is presented in Table 6.

Table 6. Biocapacity Analysis.

No	Land Use	A	YF	EQF	BC
1	Agriculture	89.138,42	1,46	2,52	326.988,90
2	Forest	106.241,34	-	1,29	137.051,33
3	Tegal/Field	144.363,10	0,82	0,46	54.615,54
4	Plantation	137.440,64	0,10	0,46	6.533,30
5	Inland water	5.553,53	1	0,37	2.054,80
6	Build up Land	80.928,78	1,46	2,52	297.753,16
	Total				824.977,03

4.5. Ecological Footprint Analysis

The following analysis of the ecological footprint, using equation (2), is presented in Table 7

Table 7. Ecological Footprint Analysis.

No	Land Use	P	YN	YF	EQF	EF
1	Agriculture	714.837,3 ton	6,74	1,46	2,52	389.069,11
2	Forest	106.241,34 ha	-	-	1,29	137.051,33
3	Tegal/Field	727.536,87 ton	5,09	0,82	0,46	54.065,74
4	Plantation	94.853,14 ton	0,64	0,10	0,46	7.048,86
5	Inland water	15.538 ton	2,80	1	0,37	2.053,23
6	Build up Land	80.928,78 ha	-	1,46	2,52	297.753,16
	Total					887.041,43

4.6. Land Carrying Capacity Analysis

The following is an analysis of the carrying capacity of land in Bali per land use category, the analysis using equation (4) is presented in Table 8.

Table 8. Land Carrying Capacity Analysis.

No	Land Use	BC	EF	DDL	STATUS
1	Agriculture	326.988,90	389.069,11	0,84	Defisit
2	Forest	137.051,33	137.051,33	1	-
3	Tegal/Field	54.615,54	54.065,74	1,01	Surplus
4	Plantation	6.533,30	7.048,86	0,92	Defisit
5	Inland water	2.054,80	2.053,23	1,00	Surplus
6	Build up Land	297.753,16	297.753,16	1	-
	Total	824.977,03	887.041,43	0,93	Defisit

4.7. Analysis of Land Carrying Capacity per Regency/City

The analysis of the carrying capacity of land in Bali per district/city also uses equation (4), the details are presented in Table 9.

Table 9. Analysis of Land Carrying Capacity per Regency/City.

No	Land Use in	BC	EF	DDL	STATUS
1	Jembrana	120.612,98	106.980,08	1,13	Surplus
2	Tabanan	116.038,83	93.925,19	1,24	Surplus
3	Badung	98.203,56	125.877,12	0,78	Defisit
4	Gianyar	89.494,13	86.276,20	1,04	Surplus
5	Klungkung	60.254,78	67.572,26	0,89	Defisit
6	Bangli	71.406,22	54.782,71	1,30	Surplus
7	Karangasem	79.705,88	80.853,97	0,99	Defisit
8	Buleleng	143.003,65	152.647,49	0,94	Defisit
9	Denpasar	55.676,01	144.416,46	0,39	Defisit
10	Bali	824.977,03	887.041,43	0,93	Defisit

4.8. Analysis of Regency/City Area Development Potential

By knowing the value of the carrying capacity of land per district/city in Bali Province, it can be seen how much population can be accommodated per district/city, or the optimal number of residents, as well as the optimal land area per district/city. Based on this information, it can be predicted in how many years the surplus regional conditions will become a deficit if the population growth rate remains as it is today, in detail presented in Table 10.

Table 10. Regional Development Potential Analysis.

No	District / City	Optimal Population (Person)	Unaccommo dated Residents (person)	Optimal Land Area (ha)	Additional Land Area (ha)	Deficit Predicti on (tahun)	Population growth (%)
1	Jembrana	313.539	-	74.665	-	18	0,68%
2	Tabanan	550.635	-	67.938	-	33	0,64%
3	Badung	522.863	147.337	53.645	9.201	-	2,36%
4	Gianyar	531.296	.	35.477	-	4	0,97%
5	Klungkung	159.705	19.395	35.325	3.411	-	0,55%
6	Bangli	296.273	-	39.956	-	44	0,60%
7	Karangasem	410.684	5.916	85.163	1.192	-	0,55%
8	Buleleng	618.865	41.735	148.066	8.763	-	0,63%
9	Denpasar	365.416	581.684	33.119	7.848	-	2,06%
10	Bali	3.962.569	374.331	619.238	48.835	-	1,21%

5. Discussion

So the results of the analysis of the carrying capacity of land in Bali based on land use categories are in deficit status, with a value of 0.93 gha/capita. The value of the carrying capacity of forest land is 1 because the available data on forest land use in Bali is in the form of non-production forest forest, therefore forest land and built-up land are assumed to be non-bioproductive land. The categories of land that are declared surplus are dry land/fields and land waters. The two land categories only have a land carrying capacity value slightly above the number one, this will not be able to support sustainable development in the future if the carrying capacity value of the land is not increased.

In general, out of 8 regencies and 1 city, only 4 (four) regencies have surplus status, namely: from the most surplus regencies, Bangli, Tabanan, Jembrana and Gianyar. Of the 5 (five) regencies/cities included in the deficit category, Denpasar City is the city with the most deficit, followed by Badung, Klungkung, Buleleng and Karangasem regencies. Overall, Bali Province is also in deficit, although some districts in Bali have a surplus. This shows that the surplus regencies in Bali have not been able to support the regencies/cities with deficit status in Bali Province.

The calculations in Table 10 refer to the formula in Table 2. Prediction of deficit is a prediction of how many years the regency/city whose land carrying capacity status is surplus turns into a deficit, if the population growth rate and land productivity conditions are assumed to remain the same as in 2019. Data on population growth are obtained from the Central Statistics Agency (BPS) of Bali Province in 2010-2019. To be able to determine the value of the predicted deficit, the population projection formula is used, namely, **Pn = P0(1+r)n**. The value of Pn is the optimal population, the value of P0 is the total population in 2019, the value of r is the average population growth rate per year, and the value of n is the number of years in the calculation area that is predicted to experience a deficit. The analysis to determine the value of the predicted deficit year (n) is to use goal seek analysis in the Microsoft Excel application. So, the essence of this deficit prediction is how many years, the time needed by districts/cities that have surplus land carrying capacity status to reach the optimal population, if the population growth rate and land carrying capacity in the region remain the same as in 2019.

In districts/cities that have a deficit or overshoot land carrying capacity status, there are several efforts that can be made to improve the land carrying capacity status, namely: Suppressing the rate of population growth; To intensify bioproductive land by utilizing relevant technology; and reduce the use of agricultural land. Intensification of bioproductive land in question is to increase the amount of production from bioproductive land by utilizing agricultural technologies, so that in the future it can increase the productivity of bioproductive land and will directly increase the value of the carrying capacity of the land.

Regencies/cities that have surplus land carrying capacity status can become suppliers or suppliers of products from bioproductive land, and can become destinations for transmigration from deficit districts/cities. This can improve the status of the land carrying capacity of the Province of Bali as a whole, and it is hoped that efforts will continue to increase the carrying capacity of the land so that it can continue to support sustainable development.

Riyadi [26] mentions that there are 3 indicators that can become a strategy in developing the region, namely, the first indicator is to increase productivity in all land categories in Bali Province, this is important because increasing productivity is the main step that must be taken if you want to increase the value of power. support land. The second indicator is efficiency, the efficiency that can be applied in the province of Bali is by intensifying bioproductive land. Intensification is meant by utilizing the right technology and human resources in cultivating a land, the productivity of the land will definitely increase. The third indicator is community participation, however the government develops strategies to develop an area, if the community does not participate or does not want to participate, these strategies will not be realized. Community participation is very important, therefore it is necessary to hold outreaches about future impacts if regional productivity is not increased.

A Ghozali and P G Aristita [2] stated that the application of the agropolitan concept can be a solution in developing rural areas without reducing the productivity of the land. The development of agropolitan areas can also be applied in various areas in the province of Bali, especially in rural areas that have extensive agricultural land. In addition to applying the agropolitan concept, implementing regulations or providing counseling so that public consumption in the use of natural resources is more circular than linear (Marganingrum, 2019) can also be a strategy in regional development in Bali Province in order to overcome the deficit status of land carrying capacity. Many efforts have been made to implement a circular pattern in public consumption, one of which is the 3R (reuse, reduce, recycle) program. This regional development strategy can be used as a reference for environmental aspects in sustainable development. For a more complete regional development strategy, it is also necessary to pay attention to the economic and social aspects which are part of the three basic aspects of sustainable development.

There are several limitations to the research on the carrying capacity of the land using the ecological footprint analysis method. First, this research only focuses on the analysis of the carrying capacity of land which refers to the production and productivity of products on bioproductive land, so the results of this study are only limited to the value of the carrying capacity of land in Bali Province, there are still many factors needed if you want to determine the value of the carrying capacity. an environment such as carbon footprint, carrying capacity of water sources, and so on. Second, the lack of data sources that can support, especially regarding exports and imports of food at the district and provincial levels, because there are still many third parties who carry out export-import activities for food products that do not collect data from the Bali Provincial Industry and Trade Office, this causes inaccuracy. data on the consumption of food resources in Bali in this study.

6. Conclusion

So, the carrying capacity of land in Bali in this analysis can be expressed in general as a deficit or overshoot. If viewed from the carrying capacity of land per district/city, the order of regions from the largest in deficit value is Denpasar City, followed by Badung, Klungkung, Buleleng, and Karangasem Regencies. The order of surplus areas from the largest to the surplus value is Bangli, Tabanan, Jembrana, and Gianyar Regencies. Even though the deficit situation does not mean that development in Bali must be stopped, because development is also a major factor in the development of a region. The development carried out must be sustainable development, by carrying out various development strategies, such as increasing productivity, efficiency, and involving community participation. As a suggestion, the government should conduct better data collection, especially in terms of export and import of food products, so that in the future it will be easier to conduct research on the carrying capacity of land based on ecological footprint calculations.

References

[1] Afni N 2016 Daya Dukung Lingkungan Kecamatan Pattalassang Kabupaten Takalar *Plano Madani* **5** 1 pp 74-85.

[2] Ghozali A and Ariastita P G 2013 Arahan Optimasi Penggunaan Lahan Melalui Pendekatan Telapak Ekologis di Kabupaten Gresik *JURNAL WILAYAH DAN LINGKUNGAN* **1** 1 pp 67-78.

[3] Rees W 1990 Sustainable Development and The Biosphere. Teilhard Studies Number 23. American Teilhard Association for the Study of Man, or: The Ecology of Sustainable Development *The Ecologist* **20** pp 18-23.

[4] Febrianto, Rahman M and Satriani 2017 *Daya Dukung Lingkungan Berbasis Ecological Footprint di Kelurahan Tamangapa Kota Makassar* Temu Ilmiah Ikatan Peneliti Lingkungan Binaan Indonesia (IPLBI) **6** Lhoksemawe: Program Studi Arsitektur, Fakultas Teknik, Universitas Malikussaleh pp 55-60.

[5] Ma B 2017 *Literatur Review on Land Carrying Capacity of the Coordinated Development Population, Resources, Environment and Economy* AIP Conference Proceedings **1890** AIP Publishing 040106.

[6] Kates R W, Parris T M and Leiserowitz A A 2005 What is Sustainable Development? Goals, Indicator, Value, and Practice *Environment Science and Policy for Sustainable Development* **47** 3 pp 8-21.

[7] Wang Y, Huang X, Gao L, Guo B, and Ma K 2018 Preliminary research on quantitative methods of water *Proc. IAHS* **379** Copernicus Publications pp 269–277.

[8] Schroll H, Andersen J and Kjærgård B 2012 Carrying Capacity: An Approach to Local Spatial Planning in Indonesia *The Journal of Transdisciplinary Environmental Studies* **11** 1 pp 27-39.

IOP Conf. Series: Earth and Environmental Science **903** (2021) 012014 doi:10.1088/1755-1315/903/1/012014

[9] Hopwood B, Mellor M and O'Brien G 2005 Sustainable Development: Mapping Different Approaches *Sustainable Development* pp 38-52.

[10] Pradhan P, Costa L, Rybski D, Lucht W and Kropp J P 2017 A Systematic Study of Sustainable Development Goal (SDG) Interactions *Earth's Future* **5** pp 1169–1179.

[11] Stafford-Smith M et al 2017 Integration: the key to implementing the Sustainable Development Goals *Sustainability Science and Implementing the Sustainable Development Goals* pp 911-919.

[12] Taiwo F J and Feyisara O O 2017 Understanding the Concept of Carrying Capacity and its Relevance to Urban and Regional Planning *Journal of Environmental Studies* **3** 1 pp 1-5.

[13] Liu Y, Zhang J, Wang S, Wang Y and Zhao A 2018 Assessment of Environmental Carrying Capacity Using Principal Component Analysis *Journal of Geoscience and Environment Protection* pp 54-65.

[14] Wei Y, Huang C, Lam P T, Sha Y and Feng A Y 2015 Using Urban-Carrying Capacity as a Benchmark for Sustainable Urban Development: An Empirical Study of Beijing *Sustainability* pp 3244-3268.

[15] Anderson C, Jovanoski Z, Towers I N and Sidhu H S 2015 A simple population model with a stochastic carrying capacity *21st International Congress on Modelling and Simulation 29 Nov to 4 Dec 2015* Gold Coast, Australia: School of Physical, Environmental and Mathematical Sciences, UNSW Canberra, Australia pp 36-42.

[16] MZ N, Soewardi K and Pratiwi N T 2015 Penentuan Daya Dukung Perairan untuk Perikanan Alami *Jurnal Ilmu Pertanian Indonesia (JIPI)* **20** 1 pp 66-71.

[17] Rusli S, Widiono S and Indriana H 2009 Tekanan Penduduk, Overshoot Ekologi Pulau Jawa, dan Masa Pemulihannya *Sodality: Jurnal Transdisiplin Sosiologi, Komunikasi, dan Ekologi Manusia* pp 77-112.

[18] Marganingrum D 2019 Jejak Ekologis Kawasan Regional Bandung *Jurnal Teknologi Lingkungan* **20** 2 pp 159-170.

[19] Goldstein B, Birkved M, Fernandez J and Hauschild M 2017 Surveying the Environmental Footprint of Urban Food Consumption *Journal of Industrial Ecology* **21** 1 pp 151-165.

[20] Wijaya I M, Prasetyo L B and Rusdiana O 2015 Evaluasi Kesesuaian dan Kemampuan Lahan Terhadap RTRW Kabupaten Kotabaru, Kalimantan Selatan *Jurnal Pengelolaan Sumberdaya Alam dan Lingkungan* **5** 2 pp 148-160.

[21] Network, G. F. (2006). *Living Planet Report*. United States: WWF Global.

[22] Lin D et al 2019 *Working Guidebook to the National Footprint and Biocapacity Accounts* Oakland: Global Footprint Network.

[23] Badan Pertanahan Negara 2020 *Luas Lahan Per Kategori Lahan di Provinsi Bali* Denpasar: Badan Pertanahan Negara.

[24] BPS Provinsi Bali 2020 *Provinsi Bali Dalam Angka 2020* Denpasar: Badan Pusat Statistik, Provinsi Bali.

[25] Disperidag Provinsi Bali 2020 *Realisasi Ekspor Impor Daerah Bali* Denpasar: Dinas Perindustrian dan Perdagangan. Provinsi Bali

[26] Riyadi D M M 2000 *Pembangunan Daerah Melalui Pengembangan Wilayah* [online] Available: https://www.bappenas.go.id/files/2913/5228/1449/bangda-bangwil1__20091008103033__2165__1.pdf

AUTHOR INDEX

Achmad, Gianoora 53
Ayu, Gusti 93
Dewi, Luh Made Ratih Ratna 134
Edhy, I G 71
Fikri, Muhammad Luthfi 41
Galford, G 1
Hardiman, G 10
Herlily 41
Hildayanti, Andi 83
Hutama, A A B Bayu Baskhara 147
Kurniawan, Kemas Ridwan 53
Mahardika, I N U 93
Mahastuti, N M M 26
Muktiwibowo, A K 147
Muliasari, I Gusti Ayu Dwi 106
Napitupulu, S S 10
Nugrahani, I G 18
Putra, I Wayan Yoga Bahagia 119
Sakti, A K 60
Salain, N R P 26
Samodra, F X T B 18
Saputra, K E 147
Saputra, Kadek Edi 93, 106, 119, 134
Siwalatri, N K A 71
Suartika, Gusti Ayu Made 71, 106, 119, 134
Suartika, Made 93
Tobing, R R 60
Tobing, RR 10
Tucker, L M 1
Wasilah 83
Widiastuti 147
Yudantini, N M 71